THE BAREFOOT SHEPHERDESS
And Women of the Dales

THE BAREFOOT SHEPHERDESS

And Women of the Dales

Yvette Huddleston
Walter Swan

Scratching Shed Publishing Ltd

First published by Scratching Shed Publishing Ltd in 2012
Registered in England & Wales No. 6588772.
Registered office:
47 Street Lane, Leeds, West Yorkshire. LS8 1AP

www.scratchingshedpublishing.co.uk

ISBN 978-0956478757

A catalogue record for this book is available from the British Library.

Typeset in Cheltenham Bold and Palatino

Printed and bound in Great Britain by
CPI Group (UK) Ltd, Croydon, CR0 4YY

This book is dedicated to everyone who lives
and works in the Yorkshire Dales

Acknowledgements

We would like to thank all the women of the Dales who appear in this book for their time and generosity in sharing their stories with us. It has been a privilege to get to know them and to learn about their lives and their work.

We are grateful to photographer Mark Pearson for the use of the image of Rev Caroline Hewlett conducting a wedding. All other photographs, unless otherwise stated, were taken by Walter Swan.

Contents

View near Kettlewell

*

Introduction

Without question, the Yorkshire Dales is one of the most beautiful regions of the United Kingdom. Much of it is governed by the Yorkshire Dales National Park and those dales that fall outside of this authority are, anyway, often classified as Areas of Outstanding Natural Beauty. It is difficult not to appreciate the attractions of the Dales: sweeping vistas, green tessellations delineated by Yorkshire's iconic dry stone walls, enchanting villages, dramatic outcrops, contours that busy the eye and, often, a combination of both rugged and lush-green scenery (generally populated by scatterings of sheep!) that is always either pleasing or inspiring, or both.

The Dales are familiar to tourists who are regularly drawn to the attractions, but tourists come and go. To live and to work in the Dales is another matter entirely, especially when compared with the cosmopolitan existence that so many of us lead - eighty percent of the British population lives in a town or city.

A painting of Swaledale by Moira Metcalfe

In writing about the Dales for several years, we have met a number of individuals who have either known no other form of existence than rural life or who have deliberately chosen to live well away from the city lights. While, superficially, it may seem easy to appreciate the appeal of quietness, the rural idyll and having few (if any) neighbours, the reality is that property in the Dales is expensive, most work is related to agriculture - which is physically challenging and of fluctuating reward - and other forms of work, even in tourism, are in short supply. Quietness is related to isolation, which requires major adjustment in how to lead one's life, particularly with no shops, or other amenities most of us take for granted, immediately available.

What this book celebrates is the way in which certain individuals - all of them women, some single, some married or divorced - have overcome the difficulties of rural living, embraced its pleasures and forged a life for themselves through their enterprise, energy and particular abilities. Meeting each of these women has been, in every case, inspiring. Living and working in the Dales has provided them with an infectious spark so that it is uplifting to be in their company. The range of their activities - shepherding, lettercarving, pulling pints, painting, treating animals, hammering on an anvil, counting grouse, preaching a sermon, serving teas and sandwiches - illustrates the variety of means by which it is possible to turn a rural idyll into a reality, tough and demanding as that can often be.

What links these special women, other than where they work and live, is a significant awareness of, and affection for, their immediate surroundings and a relationship with the landscape in which they live. With this has come a sense of values, appreciating many of the old ways while being entirely capable of adapting to, and making the best of, recent innovations. Again and again the message that comes across is that, in a community such as the Yorkshire Dales, everyone knows to lend a helping hand when it is needed, with special emphasis upon kindness, friendship and living cooperatively. That way lies harmony, both with other people and with one's entire way of life.

1

*

Alison O'Neill

The 'barefoot shepherdess' of Shacklabank Farm, near Sedbergh

"I've always been a bit of a free spirit and I've always done what I wanted to do," says Alison O'Neill sitting in the homely and welcoming kitchen diner of her farmhouse, Shacklabank, which stands in a picturesque area on the borders of Cumbria and the Yorkshire Dales National Park.

The views from the farm are magnificent - to the north are the Howgills, to the south the Lune Valley and the hills of Barbondale and from above Shacklabank the Three Peaks are clearly visible - it's an enviable location. On her farm Alison looks after around two hundred and fifty Rough Fell ewes. She would probably describe herself as a sheep farmer, she says, but has recently become known as 'the barefoot shepherdess' since she began guiding barefoot walks in the fells surrounding her home. "Walking barefoot is something I have always done," says Alison. "I remember as a child, I would either be in wellies or barefoot and I climbed most of my major peaks with my grandma barefoot. I would start walks with my boots on, but I couldn't wait to take them off."

The daughter and granddaughter of farmers, Alison grew up on her parents' farm near Sedbergh but left to go travelling when she was seventeen. "Like most teenagers I wanted to see the world and visit big cities, experience that excitement," she explains. Having been away for nearly twenty years, she returned to her roots in 1999 looking for a different adventure and determined with her then husband John to make the small tenant farm at Shacklabank into a going concern. The farm was left to the church some years previously by a farmer who specified in his will that it was to be leased 'to the poor of the parish.' "We arrived with a rucksack and sixty pounds between us," says Alison. "But I am living proof that you can survive on a tiny farm and make your dream come true."

Alison started off farming a mixture of different breeds of sheep, mainly Herdwicks and Swaledales, but now just keeps Rough Fell sheep which are quite rare, local to the area and familiar to her as they were the breed of sheep she grew up with. "They were initially bred for their wool," she says. "They are very kind sheep - kind to me, kind to each other. They are very biddable." Just two years after taking on the farm, the 2001 Foot and Mouth crisis hit and Alison and John were forced to diversify. Like many farmers at the time, they decided to offer bed and breakfast to tourists as a way of supplementing their income and they threw open the doors of the farm to visitors. Alison, who has always been a keen walker, went a step further. "That's when I decided to train as a fell guide," she says. "It all started from there really and now guiding walks is one of my main incomes, apart from my sheep; it's something I really enjoy."

Alison guides a wide range of groups or individuals on

a variety of walks which can be just for a day or over a longer period of time. "It could be a short five-mile ramble to look at wildlife or a twenty-eight mile hike over to Kirkby Stephen and then we come back and have tea and homemade cakes here or sometimes supper in the barn. I also do walks by moonlight and walks where we watch the sun coming up, or get up early just to hear the dawn chorus and then come back for breakfast."

People staying for several days can be accommodated in one of two caravans on Alison's land or in a nearby B&B run by friends. Over the summer of 2011 Alison began developing a series of walks with the National Trust in the Lake District called The Shepherdess Experience, which are day-long excursions enabling participants to get a flavour of her life. Walkers visit a lakeland farm, meet some of the people living in the stunning landscape, discover the wildlife, learn about the history of the area, what it means to be a twenty-first century shepherdess and are given the opportunity to do some wild swimming and barefoot walking.

"I really like the term 'barefoot shepherdess'," says Alison. "And it's interesting that if you look at classical paintings of shepherdesses, they are often barefoot. The term 'shepherdess' has been around for about six thousand years and I'm proud of being part of that tradition. Recently I looked up 'shepherdess' in the *Oxford English Dictionary* and the definition is 'a woman who protects sheep' or 'who guides people' which describes me quite well actually!"

Alison led her first barefoot walk with a group of people on the west coast of Cumbria. "It just really caught on," she says. "People enjoy it - they say it brings their childhood back to them. I now guide weekends and weeks where we just walk barefoot. It's better for your feet to walk barefoot and, also, I find that when you are barefoot you think more about where you are putting your feet, so you are totally focused on the walking - you don't really think about anything else. The Howgill Fells are perfect for barefoot walking because the ground is like velvet and the Yorkshire Dales hay meadows are lovely too. It's about connecting with nature - paddling through becks and mud. People always comment afterwards on how wonderful their feet feel. I especially love walking barefoot in the late evening dew - it's so cooling and refreshing. We often do wild swimming as well on barefoot walks - one of my favourite places is Uldale Force. It's something I have always done and taken for granted."

The therapeutic benefits of being in the outdoors are something Alison has experienced first-hand: after the birth of her daughter, Scarlett, she suffered from post-natal

The View eastwards from Shacklabank farm

depression and found that going walking or swimming helped greatly. "Some of the people who come on the walks are middle-aged women who perhaps have had a major change in their life and they want to find themselves again." For some it has proved to be a liberating experience, enabling them to discover something new about themselves and move on with their lives.

Alison describes the landscape around her home as 'bleak and beautiful' and herself as 'a hopeless romantic'.

"I wouldn't do this if I wasn't a romantic," she laughs. "Last winter was hard trying to feed the sheep for six weeks in the snow, I had to use my pony to get to the sheep, and we had burst pipes that flooded part of the house... so it's not all beautiful." However, she finds the richness and variety of her way of life - running the farm, tending to her

Alison O'Neill tends to her flock

sheep, introducing people to the beauty of the landscape she loves - extremely rewarding. "One day I can be cutting rot out of a sheep's foot and the next day I can be taking the comedian Ed Byrne barefoot walking on the Howgills for an article he was writing for *Great Outdoors* magazine. It's a great life."

Now amicably separated from her husband, John, Alison lives alone at Shacklabank with Scarlett - although John lives nearby, still helps her out on the farm occasionally and sees their daughter regularly. Being a single working mother running a farm is not always easy and Alison is honest about the fact that sometimes it can be quite a challenge, but for her the positives far outweigh the negatives to make the hardships worthwhile. "I had to pull a dead sheep out of the river the other day - I literally had to drag it out, which was really tough but you just get on with it. On the other hand, every morning I walk with my dogs up to Fox's Pulpit where the founder of the Quakers George Fox gave a sermon. It's about a mile from the farm and it's my treat. I always think how lucky I am to start my working day like that instead of sitting in an office. I'm no different from anyone else, it's just that I am doing something different. I'm just an ordinary working-class woman, living on a rented farm on a hill in the Howgills and making a living."

That is a modest self-assessment since Alison has won numerous awards for her initiative including the *Country Living* Magazine Women in Rural Enterprise Award 2006 and the North West Tourism Experience of the Year in 2008. It is partly as a result of this that she is in demand as an inspirational speaker. "My talks are really about how you don't need lots of money to make your dream come true," she says. "I want to show people that my way of life is achievable. My dream was to farm and have this life. Everybody thought it was impossible but, if you are determined enough, you can do it."

Her audiences can range from interested parties such as the Yorkshire Dales Society and the Countryside Alliance, to ladies' lunch groups and farmers' discussion societies. "I find the talks really easy actually. It's a lot like therapy - talking through things that have happened. It is sometimes the everyday things that people want to hear about. I used to leave out all the struggles I had - like having no money and sheep dying and so on - but some of the people I speak to would be interested in that." She says that, although she enjoys presenting her talks to all kinds of people, speaking to other farmers can be very satisfying from a personal point of view. "Recently I spoke to a group of landowners and I talked about some of the difficult aspects of farming and keeping tups and sheep. They appreciated my honesty -

they wanted to hear the truth. I think that was really the first time that I felt I could talk completely openly about how hard it can be."

None of the hardships seem to faze Alison, however, as her fundamental belief and joy in her way of life is quite apparent. "I'm very independent and I think as you get older you begin to feel better about yourself. I don't have to explain myself to anyone. There are downsides to being single, sometimes in the evenings I would like somebody to talk to, but generally I really like it." She is clearly someone who is comfortable in her own company but as a warm, friendly and naturally gregarious person, her work outside of the farm, guiding walks and giving talks, enables her to satisfy her sociable side. "I love my animals and I love my life but I also love people and I do love doing my walks and talks and meeting people. I'm interested in people; everybody has something exciting about them."

Her main business, however, is still her sheep and she regards that as both a privilege and a responsibilty. Sheep are notoriously prone to disease - "my dad always says 'sheep are born to die'," laughs Alison - so there are inevitably times when Alison will lose sheep, sometimes for no immediately apparent reason. "When I first came here I used to get really upset when a sheep died - I still do. I try to be hard but I still get upset. It's not just that they have gone - but on a practical level, it is also part of your income." If a sheep is obviously sick, Alison will make sure that it is not alone when it dies. "I will put her in a barn with some of her friends. Sheep really do have a strong connection with each other. The family bond lasts - yowes will recognise their lambs even when they are grown and go and stand next to them on the fell. It's sad when a sheep dies, but then lambing time will come round again - so it's all part of the cycle of life."

Alison enjoys a rare relaxing moment

She says that lambing time is her favourite time of year because it feels like a period of rebirth and renewal. "I love it: it's a sign of Spring. It's also my busiest time - I am always on the go, checking on the lambs, feeding those that need extra milk with bottles and keeping an eye on all the yowes." She has a shepherd's hut - similar to a traditional horse-drawn gipsy caravan - and will take it up onto the fells with her pony, Sunny, ("I saw him being born as the sun was rising. He's my quad bike") during lambing time. "It's good to be nearby in case the sheep need me. I love my sheep."

Her bond with the animals is clear. Seeing her striding out on the hills near her farm, with the shepherd's crook that her father made for her many years ago out of an antler that he found on the Howgills and a hazel branch ("It means a lot to me"), rounding up the sheep with the help of her dog, Joss, it's obvious that she is a natural countrywoman who is comfortable in her landscape and very much at one with nature. "I was obsessed with animals and nature as a child," she says. "And nothing much has changed. I love the fact that just looking out of my window in the morning I can see lots of varieties of birds including buzzards - very useful during lambing time because they keep the fields clean by picking up the afterbirth. Then there are the roe deer, badgers, otters and voles."

Living within the landscape and connecting with it is all part of the pattern of Alison's life. "I never wear a watch; I function with the seasons," she says. "In the summer I get up every morning and go for a swim in the River Lune down at the bottom of the valley - it's crystal clear and there are often kingfishers flying around just above my head. I love sitting and watching my bird-table while I have my breakfast. One morning I woke up to see woodpeckers on the window ledge - it makes everything worthwhile. When

I am leading my walks I will stop after lunch and we have a half hour of quiet time - when I shut up for a while! - and I tell people to just listen to the sounds of nature. Nature never ceases to surprise me."

As well as her talks and walks, Alison has recently developed another diversification - into clothing design. Alison has her own very distinctive style - her preferred working and walking clothes are a tweed skirt, waistcoat and jacket, with thick woollen socks and walking boots - and she has started her own line of tweed skirts and jackets, all to her own design. "I love tweed and have worn it ever since I was a little girl," she says. "My dad and my grandad always wore tweed jackets. I remember my grandad used to give me his jacket to keep me warm out on the hills when I was a child. Everybody wore tweed when I was growing up. I was twelve when I got my first Harris tweed jacket, a secondhand old hacking jacket from an auntie in Sunderland. I had a grandma who always wore tweed skirts and Fair Isle jumpers; she had tweed skirts in every colour. She was a real inspiration for me - she was an amazing woman and a bit of a free spirit. She loved nothing better than packing a little rucksack and heading off to walk in the Dales for the weekend."

Inspired by the memory of her grandmother, Maggie Winn, and by her own affection for Harris tweed, Alison set to work. "I did some designs and then I found a friend living nearby who would make them up for me. I now have two jacket designs and several styles of skirt." Designing the tweed outfits fulfils another aspect of Alison's personality. "When I was seventeen I wanted to go and study Fine Art but instead I went travelling. That's my only regret, really, that I didn't go and do something creative, but I have never stopped drawing and now I am doing my tweed designs." The Shepherdess Range includes four different skirt designs,

a waistcoat and two jackets and Alison is planning to develop further ranges.

The tweed is woven by hand in the Outer Hebrides and the garments created in Cumbria; Alison's long-term aim is to create tweed out of the wool from her own sheep. She has spent a lot of time on the island of Harris and feels an empathy with the islanders and their simple way of life. "I have a friend, Mary Ann, who lives on a croft up there and in many ways we lead a similar sort of life - except the weather is even worse up there!" Sales of the tweed designs have been going well and Alison has had orders in from as far afield as France and Italy. "At the moment I am trying to put together a three-year business plan," she says. "All three areas that I have developed - Walks, Talks and Tweeds - have started picking up and I don't really want to be much busier. I always want to be a shepherdess, but the reality is that I have to diversify if I want to keep this lovely life and provide for Scarlett."

Proud of her roots ("I'm a Yorkshire lass through and through," she says), Alison feels a profound connection with the landscape and people of the Dales. "One of the questions you hear around here is 'How's ta bred?' and if I say 'I'm a Winn from Sedbergh', people will know. It's like having a pedigree! I will explain that my family was originally from the Dales - on my father's side they were lead miners who were given a little piece of land near Horton-in-Ribblesdale and began farming." Her mother's side of the family are originally from the North-East who came down to Yorkshire to work in service. "The girls from the North-East were hard-working and it meant that the Yorkshire farmers got to marry women that they weren't related to!"

Alison cites Hannah Hauxwell and Alfred Wainwright as her heroes, both close friends of the Winn family. She admires Hannah's resilience and strength of character and Wainwright's ability to write about and illustrate so evocatively the landscape she loves. It's clear that the way of life she has chosen, though sometimes difficult, brings her a great sense of personal satisfaction and achievement - quite deservedly so. "I feel joy," she says. "I still feel like I'm a fourteen-year-old even though it won't be that long before I'm fifty. I'm doing what I love and living my life how I want to live it. I make a living and provide for Scarlett, but nothing is set; I just blow with the wind."

Walks, Talks and Tweeds at Shacklabank Farm
www.shacklabank.co.uk

2

*

Amy Lucas

Gamekeeper, Cotter Dale

Amy Lucas, who is in her early twenties, is a gamekeeper in Cotter Dale, an offshoot valley at the western end of Wensleydale, beyond Hawes. The hamlet she lives in, Cotterdale, is a small, sociable community.

"Everyone is very friendly," she says. "There are four gamekeepers' cottages and then four or five houses that have people living in them all the time. They are all country people and local - related to families living nearby. There are some second homes too and holiday cottages that are always quite busy but especially in the summer. The Pennine Way path is up on the tops of the hills so, in summertime, we get people passing through all day. Some people will stop and talk to you - they seem genuinely interested in what we are doing."

What Amy does is to work on the Simonstone estate, which includes Abbotside Moor, and her main role is to ensure that the habitat encourages grouse to flourish on the moor. When we met her most recently, in early May, it was a particularly busy time of year. "All the vermin are having young. We are on the lookout for foxes because a pair with cubs can do a lot of damage - they eat the grouse eggs and chicks." There is no escaping the fact that, if grouse numbers are to be maintained in preparation for the Glorious Twelfth and the shooting season, a significant aspect of the local economy, then the grouse have to be protected from predators such as crows, stoats, weasels, rats and foxes. Grazing sheep also pose an indirect threat to grouse - they eat the heather that is an essential part of the grouse's diet and habitat - so the estate has reached an agreement with the local farmer to restrict sheep in the dale to lower lying areas.

Weather is also a factor. "When the grouse start to hatch you really want it to be dry. Last year we lost a lot of chicks because it rained so much that they drowned - we were just picking them up off the ground." As ground-nesting birds, they are extremely susceptible both to predators and to flooding but Amy, along with her immediate boss, head gamekeeper Paul Starsmore, were very happy with last year's season: "one hundred and twenty brace was our record day," says Amy, "and on most of the shoot days we did have perfect weather, which makes a change!"

When standing on the upper reaches of Cotter Dale, you can see for many miles in every direction. When the wind isn't blowing, it's possible to sense the kind of silence that few human beings in this country are ever able to experience.

Occasionally, the guttural and faintly comical honking of the male red grouse will intrude upon the serenity of the scene but it is the tranquillity of the location that impresses. There are generally between six and ten shoots planned each season when that tranquillity will be broken, with the

number of birds shot strictly regulated by the estate management and head gamekeeper.

A self-possessed, practical and self-contained young woman, Amy is clearly very passionate about, and engaged in, her work. Having been in her role for nearly two and a half years she seems to be fully in her stride in handling what is expected of her. "I'm really enjoying the work. I'm just starting my third year here so I've been here long enough to know exactly what I am doing," says Amy. "A lot of people start to look for another job after about two years but I'm happy where I am and I'm not looking to move on - I like it here and I get on with everybody. I'm still the only woman and I don't think that will change. The first year you are picking it up," she continues, "learning about what to do and when to do it. It's the kind of job where my boss isn't

Amy with one of her spaniels

telling me what to do every day - I'm left to get on with it and I like that. It helps if you get on with your boss too."

Interestingly, Amy's great-grandfather had been a gamekeeper in Lancashire but it is her instinctive love of being out in the open air that most attracted her to this kind of work. "When I was at school I did a work placement for a week helping out on a shoot at Helbeck Hall near Brough," Amy recalls, "and I loved it. I did one year of sixth form, studying Biology, Chemistry and English, but I didn't particularly enjoy it. I've always been interested in the countryside and for me this way of life is better than reading a book. So, I took a two-year national diploma course in gamekeeping and countryside management at Newton Rigg College in Penrith. The course was a good mixture of practical outdoors work and classroom work."

It was obvious at college that it was relatively unusual for a young woman to be preparing for a life as a gamekeeper. "There was one other girl on the course," says Amy, "but she was more interested in countryside management and you can go on to university with that and study further; she hasn't gone into gamekeeping. I think the countryside management side of things appeals more to women than the work I do." After she had completed her course, Amy worked in Scotland for a few months on the Glenshee estate near Rannock Moor. Later, she was based at a game farm near Appleby for a while and she found some beating work on shoots near Reeth. "I got to know quite a few gamekeepers that way, I told them I was looking for work and I heard that Paul was looking for an underkeeper here. It just worked out really well. I've got a nice little cottage that comes with the job, then went on contract, and learned a lot in a short time. Paul is a great boss and a really good teacher."

Amy's present hours are not too arduous for countryside

work - "you are up and out by eight o'clock, then back for lunch and out again, back by about half past five in the evening. That depends on what you are doing, though - the other day I was up at four-thirty in the morning to sit out and see if there was a fox. That time of the day is the best time to catch them unawares. Then about seven in the evening we might go out again for a couple of hours, weather permitting," says Amy, "looking out for pests and predators." She also goes out during the night, mainly seeking out rabbits by using a large lamp attached to a land rover to spot the prey. Wherever Amy goes, her shotgun goes with her - and with it comes responsibility. "I have a shotgun licence and you need a firearms licence if you have any rifles, which I do," she says. "You obviously have to be very careful with your gun. You can't leave it in your vehicle. If you lose it or it gets stolen, then it's your responsibility - then you lose your licence and your job."

For Amy part of the appeal of her work is that, apart from certain tasks that need to be carried out every day, she is not tied to a monotonous routine. "No two days are the same," she says. "The only thing you have to do is to check your traps every day. My job falls into three areas - pest and predator control, habitat management and making sure the grouse have access to grit to aid digestion. We are basically protecting the grouse. We have different types of traps but the main predators are stoats and weasels, rats and crows. They all take the grouse eggs and the crows will sometimes take the young chicks."

Amy has acquired a variety of means for recognising when the grouse might be under threat and uses all her senses to detect any sign of danger. Being alert to the sounds of the countryside, for example, can provide helpful information.

"One of the useful things about lapwings and curlews is

that they give out a very distinctive distress call if there are vermin about. The curlew has quite a pleasant call normally so it's obvious when there is something not right. The other thing we do is go tracking - especially up on the tops where the peat is wet because it's easy to spot tracks in the peat. It's a good indicator. If you do see something, you might go out lamping later or in the morning. We do a lot of our tracking in the snow which is great - except when there is a wind and the tracks are blown over." When tracking she is alert to every potential sign, even sniffing for the distinctive smell of fox urine or locating fox scat (faeces). If there is evidence of telltale footprints or other indications, then Amy is obliged to undergo a late night vigil with rifle to hand.

Though there are four gamekeepers working in Cotter Dale, two of them concentrate on maintaining pheasant stocks - Amy much prefers working with the "more

A view of Cotter Dale

intelligent" grouse which, she explains, are wild birds who choose to remain on the moor if the conditions are right. "Because they are a wild bird you have to work that bit harder. They are very independent. If they are there it's because of what you have been doing, which is satisfying," says Amy. "They are really intelligent and resourceful birds; they manage to survive in really bad weather conditions. I've seen male grouse pretend to be injured, for example, if you go near a nesting hen, to distract you."

Amy's only immediate co-worker is Paul, the head gamekeeper, and most of the time she is on her own. "You are alone almost all the time - nine times out of ten you will be working on your own unless there are specific jobs to do. However, I am happy in my own company and I like being outside. You can sometimes get sick of it when it's been wet all week and you might think, 'I wish I worked in an office,' but most of the time it's really good to be outside. I couldn't think of anything else I would want to do. Things can change but I can't imagine not doing this work. It's more of a lifestyle than a job - and it's a job for life. I've always been happy doing my own thing." Looking to the future and the prospect of getting married and having children, Amy ponders the possibilities: "I think if I ever had children then I would have to give up my job as a gamekeeper because it isn't the kind of work that fits in easily with family life. But that's a long way off..." Equally, she is aware that you need to maintain a balance. "I know some gamekeepers who are obsessed by their work and that's not good either."

The working year is geared towards the shooting season, a time which provides Amy with some of her greatest satisfaction in her work. "What makes it all worthwhile is on a shoot day seeing everyone enjoying themselves and hearing the 'guns' talking about how good it was." ('Guns' is the term used by gamekeepers to describe those who make up the shooting party, normally between eight and ten in number). To assist the 'guns', Amy explains, "there are ten beaters and eight flankers (usually from the older generation because there is not so much walking involved). They are mainly people from local farming families - the beaters are young lads on their school holidays." Amy derived her interest in the sport from her father. "He's a plumber," says Amy, "but he has always been interested in shooting. I started shooting clay pigeons with my dad when I was about twelve. Then I would go beating at different places." Progressing to work on the ten thousand acre Simonstone estate seems a natural development.

An important aspect of Amy's job is to burn the heather on the moorland in order to manage the habitat for the grouse.

"We have to burn so that there is heather on the moor at all different stages of its growth because that's how the grouse like it," she explains. "They eat the heather shoots, seeds and flowers. We set lots of small fires and then use 'flappers' to control them. We have to do burning when the weather is just right - not too wet and not too dry. So, we just do it when we can. Burning is generally done from October to April. The grouse nest in the heather and maintaining the heather helps other birds too, such as lapwings and curlews. During the nesting season, in April and May, we stay away from the heather and also when it's really wet - we don't want to disturb the young grouse."

This is the time when it's also important to encourage walkers to stay on footpaths and to discourage untrained dogs from being loose. Dogs, however, are an essential assistance to gamekeepers in their working day throughout the year. "You need to have dogs to do this job," says Amy. "I have three spaniels and one of them has just had five puppies. I will keep two of them and I'm selling the others.

The dogs really help out on count days. We cover as much ground as we can to count the breeding pairs in March and the broods in July. I take the dogs out most days especially if I am training one - sometimes it's just nice to have a dog with you for companionship."

The moors can be threatening and dangerous if you don't know the terrain and in extremes of weather, but the gamekeepers are safety-conscious. "You always let someone know where you are going and we usually have our radios with us so we can contact somebody. I've never yet had a situation where I've had to call anyone for help. However, you do have to be careful not to tip off your quad bike."

The quad bike is a major blessing for Amy, as is her Land Rover, but walking still makes up a good part of her working day, especially when checking traps for rats and weasels. "You have to keep your quad bike in good condition because you wouldn't be able to do your job without it," says Amy. "I do a lot of walking too. I need the bike to get up to where the traps are but then I have to walk to get to them. I think gamekeepers in the past must have been very fit!" Only once has Amy felt a little vulnerable while out checking her traps. "I came up onto the moor one day when the mists came down suddenly and, for a short time, I didn't know where I was so I just had to follow my own tracks back." The weather can change so suddenly in these parts that Amy always ensures that she is properly equipped - and that means dressing appropriately. "You have to wear clothes that will cope with all four seasons in one day up here," she laughs.

Amy's enjoyment of her work is palpable - she lives and breathes it. The best aspect of it, for her, is that her workplace, though wild and rugged, is part of the beautiful Yorkshire Dales. "What makes my job enjoyable," she says, "is being outside - and the freedom. It's very flexible, there

Amy checking one of her traps

are no daily deadlines. On shoot days, if things aren't going well because of the weather, it can admittedly be a bit stressful - but it's really Paul who gets stressed about that, not me!" Another appealing aspect of where she lives is contact with wildlife which doesn't represent a threat to the grouse. "I get to see roe deer, hares and, sometimes, some

Amy Lucas is committed to her largely solitary job

interesting lizards. We get quite a lot of red squirrels round here too which is nice - they don't cause us any trouble."

Though Cotter Dale is no great distance from Hawes, Amy is not someone who feels the need to socialise on a regular basis in her free time which, perhaps, explains her contentment in a largely solitary job. "I have never been one for going out much; I've always been like that. I like it quiet but, having said that, with all the keepers living in the village there is always somebody to talk to. We usually keep to ourselves though. If there is a social event at Simonstone

Hall during the shooting season then we all go along. Quite often in the evenings you have work to do anyway - going out lamping or looking for foxes. After a shoot day we will go down to one of the pubs in Hawes - and there are gamekeepers' dinners at the end of the season." More usually, though, Amy will spend time in the evenings on her computer or occasionally watching television, though she has recently purchased a mountain bike and is looking to make good use of it within the dale and up on the moors, especially during the summer.

Weekends, though, are for catching up with her family. "They are not far from here - only about half an hour's drive away." Amy is the eldest of four siblings - she has two brothers, aged twenty-one and six, and a sister aged twelve. "One of my brothers, the twenty-one-year-old, works on the Helbeck Estate as a keeper of pheasants and partridges. He was training to be a mechanic but I think seeing me doing this job encouraged him to think that he would like to try this kind of work." Although independent and self-sufficient, Amy is close to her father and mother and appreciative of their support. "My mum has always been around for me - and I wouldn't be who I am today if it wasn't for my dad."

When young people are sometimes castigated in the modern era (mostly unfairly) for selfish attitudes and lack of respect for previous generations, and little sense of direction, it is extremely refreshing to get to know someone like Amy. She has a genuine love and enthusiasm for her work and her workplace, and appreciates how much she has benefited from what her mentors have taught her. She appears to have found contentment young and says of her life and work: "This is what I'll always want to do. I don't see it as a job - it's a way of life."

3

*

Annabelle Bradley

Artisan Blacksmith, Malhamdale

Annabelle Bradley, originally from Bradford, lives with her husband Nick and their two primary school age daughters, Millie and Hatty in Malham where, since 2007, she has been the village blacksmith. Taking on the smithy at Malham involved two major life decisions for Annabelle - the kind which not everyone is courageous enough to make.

One was to decide exactly where to live when the option to move house arose, involving the possibility of abandoning town life in Bingley, near Bradford, and her straightforward commute to work and, instead, take up country living in one of Yorkshire's most visited villages. The other decision, having by then spent four or five years in Malham, was to give up her regular job (and salary) as a tax accountant to work at the Malham smithy full time when that opportunity became available. "The smithy had been empty for a couple of years and the church advertised for

Annabelle is a busy blacksmith in Malham

craft people who might be able to use the space," says Annabelle. While the church had advertised the space as a workshop for any craft, the wardens had hoped it would be retained as a smithy as stated in the original bequest when the property was donated to the church by a previous artist and blacksmith, Bill Wild.

"It's very good he did leave it to the church because it has ensured that it has been kept in use as a smithy," says Annabelle. This was a wise provision, without which the forge might have been used for some entirely different function, or even left to ruin, thereby losing the village a significant slice of its heritage forever. "Parts of the building are estimated to be around three hundred years old," says Annabelle. "It would originally have formed a section of the outbuildings or barn for the farmhouse opposite." The

A view of Malham Cove

farmhouse, which is now the River House hotel, dates back to 1664 and was originally called Aireview.

The second decision was very much influenced by the first. Having worked for Filtronic (who are based in Saltaire and specialise in wireless communication equipment) since 1995, and continuing to do so after buying a cottage near the forge, Annabelle found her commute from Malham every day had become a major undertaking. "It took me well over an hour each way and there was always a lot of traffic," she explains. However, Millie and Hatty's needs clearly influenced how Annabelle wanted to organise her time, and theirs. "Before my youngest daughter started school I used to drop her off at nursery in Skipton on the way to work but when she started school she had to do breakfast club and the after school club - they were both doing that. They were so young, it didn't seem right for them to be doing such long days."

Switching from the comfortable and lucrative routine of accountancy to working as a blacksmith was, however, less of a leap in the dark given that, for some considerable time, Annabelle's hobby had been jewellery-making. "It was based on commissions and family and friends and then it got to the point where I was so busy that I thought, 'I could do this full-time.' I was doing beadwork as well; it was all self-taught. I bought all the equipment and played with it, experimenting."

Essentially, then, could Annabelle justify the loss to the family income of her prestigious job with Filtronic weighed against the uncertainties of a career based upon an artistic pastime? "In 2006 I was getting closer to making my decision," she says. "The way the company I was working for was developing, my role was likely to be coming to an end within the next two or three years. That helped me to decide. It's not an easy decision to make but, in the long term, you are making your life better even if it's a hard choice to make in the first place." She eventually started in the forge in February 2007 with another blacksmith, David Crane, and the two of them developed their skills together. David had been on a basic blacksmithing course and he showed Annabelle some of the techniques he had learned. They spent some time clearing out the smithy and making it into a practical working environment and held their first Open Day in April 2007. "It was a bit nerve-wracking," says Annabelle. "Would people come and would they want to buy anything? But it was ok and the tourism in Malham in the summer helps."

For the first six months Annabelle still worked part-time at the offices in Saltaire from Monday to Wednesday, and then worked from Thursday to Sunday at the forge. When she realised that she really could make a go of running the smithy, it was time to leave tax accountancy behind and devote herself full-time to her new profession. "Because I have two young children, what I do now fits in better round our family life - not trying to fit the children round our work. Now, the children get off the school bus in the village and come in to the forge. It has worked really well for us as a family."

Annabelle's husband, Nick, works in the National Park selling and marketing building materials to various projects and architects in the North of England. "He spends most of his day driving round the Dales - and his is a family business working with his dad and his brother so, as far as our family is concerned, it's ideal." Between them Annabelle and Nick have solved what is one of the greatest obstacles for relatively young people attempting to raise a family in

the Dales - finding regular work without having to spend hours each day commuting to the nearest large town or city.

"Malham has always been a very special place for my husband Nick and I because we came here on our first date for a picnic and we always came back here every year. When we decided we were going to move, back in 2001, we weren't necessarily thinking of Malham, we just put our house on the market and thought we would sell it first and then see what happened. We had always thought that we might move to Malham in about ten or twenty years time and that it would be our final destination. But when we saw a cottage in Malham advertised, we just had to have a look at it. The people who owned it made us feel really welcome and at home. We went for a drink at the pub and thought about it - and decided to go for it! That was in the middle of the Foot and Mouth crisis, so no-one else was coming in to the Dale at that time, really."

In taking on the lease for the forge, Annabelle is maintaining a link in a chain of smiths working in Malham. "The blacksmith, Bill Wild, who was at the smithy from 1946 until his death in 1984, had no children, and it was he who left the smithy to the church - so it came with all the equipment, which is great. He made the big capital investment," says Annabelle. "There were three blacksmiths after Bill - Donald Rawson was here for about ten years until his death, then John Clements (from Kilnsey) and his son, who is now a blacksmith in Settle. And then Phil Mason who left in 2005. They were all traditional artisan blacksmiths." Seeing Annabelle at work in the forge is thoroughly absorbing. Both her skill and her enjoyment in her work are very apparent - it seems almost as if she was born to be doing what she does. She is also more than happy for members of the public to call in and see her in action with hammer and anvil.

Annabelle's ornate handrail on the bridge in Malham

Annabelle's first creations which were offered for sale included coat hooks, pokers, candle sconces and fireplace companion sets. As her expertise developed and became better known, she was commissioned to produce particular pieces of metal craftsmanship. Television viewers of the first *The Dales* series on ITV (hosted by Ade Edmondson) will have seen the process by which she made three chandeliers for the Buck Inn, the pub and hotel that stands close to the village green and to the smithy. More commissions have followed.

"The Parish Council got a grant to replace the old wooden footbridge in the village with a more traditional style stone clapper bridge and they asked me to make the decorative wrought iron handrail. That was very satisfying but it has proved to be my hardest job so far. It wasn't until doing that commission that I found the work physically difficult. It did take it out of me but it was great to do something that is now part of the community and it's something that will endure. I like to think that one day in the future some children will be playing on the bridge and say, 'My great-grandma made this.'"

Despite the inevitable initial difficulties of setting up a new business, Annabelle did not take long to feel that she had made the right decision to make her craft her profession. "It is extremely satisfying that people like my work, have bought it and then taken it into their home - that feels like a real privilege." The knowledge that there is a market for what she crafts has boosted her self-belief and allowed her to develop as a blacksmith. "Over the past five years my style has changed," she acknowledges. "This past year has been extremely busy and now I have to refocus and ask myself, 'What is it I want to do?' I don't want to just make things that I know people will buy. I want to make things that I am proud of and that can be recognised as an artistic piece. I feel a lot more confident - I know what I want to achieve and I'm confident to take on the challenge. People now come to ask me for advice - and I like that. I am definitely more adept than I was and I can make things quicker than I used to but I don't want to become a production line. I want each piece to be unique and, anyway, no two things I make are ever exactly the same." For many, that is precisely the appeal of a hand-crafted piece - it is distinct and subtly different from any other.

Developing her expertise is a great deal easier for Annabelle than it would once have been for anyone attempting to acquire new skills. "There is so much information on the internet now which is such a great tool. When I first started I got Peter Parkinson's *Guide to Blacksmithing* on DVD - it gives you a great insight into various techniques. I am basically self-taught. The best way to learn is as an apprentice but if that's not available then the next best way is to just get on and learn it yourself by doing it. The nearest course for me would be in Preston and it wasn't practical to try and do that and run a business at the same time."

However, there are other means by which Annabelle can continue to build on her knowledge at the same time as passing it on to others, particularly the general public. For the past few years she has been taking part in the Great Yorkshire Show, invited to perform demonstrations by the Worshipful Company of Blacksmiths. "When I was at the Show," says Annabelle, "I was with eight blacksmiths, all men, and some of them have been practising their craft for twenty or thirty years. It was really good to be able to speak to them and benefit from their knowledge. It was very pleasing to see my work alongside theirs. What was interesting was that, if you looked at the displays, you could see which pieces were made by a woman - there is a distinct

difference. The men tend to use heavier metals but it is also that as a woman I tend to bring out the softness and delicacy of the metal. What was really exciting was that I won two awards as well!" Annabelle's softer and more delicate approach nevertheless obliges her to hammer out heated metal on the anvil - and it's demanding, physical work. "In the first few months I found that my wrists were very tired from keeping them strong when you are hammering but then after that it was fine. I do find that if I go on holiday for two weeks it takes a while to get going again."

In the past, and before her appearance on television, approximately forty per cent of Annabelle's work was commissioned; the rest was bought by passing trade. With her fame having spread, it's now seventy per cent commissioned work. "With commissions I make things to people's specifications," Annabelle explains. "If someone brings me a drawing, I do try and follow it, although I don't get as much enjoyment out of that. It's better when they trust me to do things. I have had commissions from all over - there are so many people who come through Malham. I have even had some from overseas: I've sent coat hooks out to Chile and a door knocker to Finland. This year was the first time that I didn't close in January and February because my order book was so full. From October onwards there were lots of orders for Christmas - and some of them I couldn't complete but people are very understanding when they are having something specially made. I've currently got a three month backlog!" Her popular items at the moment include curtain poles, fire baskets (she recently made one for one of the pubs in the village, The Listers Arms) and fireside tools. Occasionally, Annabelle has to turn down commissions as a matter of principle: "I have had requests for re-enactment swords and such like, but I don't do anything that might be categorised as weaponry."

Apart from the obvious appeal of living in a friendly village with beautiful countryside almost literally on the doorstep, Annabelle appreciates that the rural landscape that surrounds her has strongly affected her artistic work. "Rams horns and shepherd's crooks are both an influence, and the tapers and twists that I apply to the objects I make are connected with the water around here - that is all very much the influence of the dale." As well as what she derives from the landscape, Annabelle's ideas are sometimes purely a matter of inspiration. "I often just wake up with an idea in the middle of the night! But I don't often put my designs down on paper - it always ends up changing when you are working with metal, anyway."

It now seems almost impossible to imagine Malham without Annabelle working in the smithy, but there were certain obstacles to overcome before she began to feel settled. "When I first moved here, I was probably about two weeks pregnant at the time - although I didn't know it then. It was November and I was coming home from work in the dark and I found it quite difficult. But then when Spring came, suddenly this wonderful world opened up - and I've never looked back. Nick and I made a conscious effort to get really involved in village life: in the first year we joined lots of local committees and we are currently involved in the PTA, school governors and the parish council. Malham is a fantastic place to bring up kids - they have the freedom of the village. It's wonderful that the children have the kind of freedom I had when I was younger."

Annabelle is grateful that Malham is a thriving Dales village with plenty going on. Most of the properties are lived in full-time by local people. "There aren't that many second home owners," says Annabelle. "There are some holiday lets, but they keep the village busy with plenty of people coming through and that's good for local businesses. I love living here and I can sometimes not step out of the village for six weeks at a time. Asda and Tesco will deliver here - the nearest shop is the Co-op in Gargrave, so you generally call on neighbours before you go to the shop. There are lots of hens in the village so there are always plenty of eggs!"

The friendliness of the village is renowned. "When my dad comes up for the day he says he wonders how I ever manage to get anything done because people keep dropping by to the smithy. It's nice that the smithy is used as a centre for a chat and I really like that people who live in the Dale bring visitors or family in and introduce me as 'the village blacksmith'. I also love the fact that my daughters don't find it unusual that I'm a blacksmith. My eldest said the other day that when she grows up she'd like to be either a blacksmith like mummy - or a ballet dancer!" Both girls have already been introduced to working with the anvil and have created their own metalwork art pieces.

Family life strongly influenced Annabelle in her decision to take on her distinctive profession and it remains a key part of her day-to-day life, including taking a well-earned break from work. "We have a campervan which we use for family holidays. I have to be here at the smithy over the summer, really, because that's when the tourists are here and I get the most passing trade. Because the smithy belongs to the church we don't open on Sunday mornings, but I am here in the afternoons. We often go camping overnight even if it's just a mile down the road - and then I can get back to open up the smithy on Sunday afternoon. As a family it's nice for us all to switch off together."

Annabelle's story, like that of so many women in the Dales, serves as an inspiration for others to have the courage to change their lives in a positive way.

"People often say to me, 'If only I could do something like that.' Well, you can! You just have to do it. I was lucky that my family supported me one hundred per cent and convinced me that we could do without the security of my regular salary. Everyone was positive that this was the right thing to do." It's obvious seeing Annabelle at work and hearing her speak about her life in Malham that she has certainly done the right thing for herself and her family. She sees herself continuing to be the village blacksmith for a long time to come, with the smithy as an integral part of family life. "I can't see a time when I would be prepared to hand back the keys to the building. I feel proud being part of the village's blacksmithing heritage."

www.malhamsmithyonline.co.uk

4

*

The Reverend Caroline Hewlett

Vicar of Swaledale with Arkengarthdale

Caroline Hewlett is the Vicar of Swaledale with Arkengarthdale (or as her Facebook page jokily has it 'Vicar of Dibley with Ambridge and Camberwick Green') and lives with her husband Alex in the vicarage on the outskirts of Reeth. She almost certainly drives more miles attending to her parishioners than most priests anywhere in the country - she can cover around fifty miles on an average Sunday. Through necessity, she has a four-wheel drive vehicle ("When I first arrived I started off with a little saloon car, but quickly realised that wasn't going to work," she says) and shortly after taking up her post in 2006 took an off-road driving course to help her cope with the more challenging terrain of some parts of her parish which extends into the remote stretches of the northern Dales.

Her off-road training has come in very handy especially during one snowy Christmas when she was due to attend a joint Anglican/Methodist carol service at Arkengarthdale

Methodist Church. The service alternates between the two churches and the minister of the church usually leads the service. "The road up through Arkengarthdale was very bad," says Caroline. "You could hardly tell the road from the moor but, using my new off-road skills, I carefully made my way to the church." When Caroline arrived she was greeted by the organist who seemed very relieved to see her, explaining that none of the other ministers could get there and that Caroline would have to lead the service. "So, at two minutes' notice, with no robes or books, and with one service sheet between us - the rest were snowed in at someone's house - I led a carol service for around seventy people who had walked there from the village in the snow." On her way home she stopped to help a delivery van driver who was stuck in a snowdrift on the moor. "I was only able to stop because I knew how to get going again in those conditions. It was about minus seven that evening!"

Ordained in 2001, Caroline then worked as a curate at St George's in inner-city Leeds, which does a huge amount of valuable work with homeless and disadvantaged people, and got her first taste of rural ministry in Boroughbridge before a three-month secondment to Swaledale in 2004. She deliberately chose to come on placement in winter in order to see Swaledale at its most challenging - and loved it so much that she decided to return two years later, applying for the post of vicar when it became available. "You are very much a part of the community here," says Caroline. "There was a gear change after year five - I stopped being the new vicar and I feel that I am part of the place. You are expected to attend as many of the social events as you can and everyone knows each other - it would be very hard to remain anonymous. I can't go anywhere without being recognised: it takes a while to get used to the public nature of the job, but I like that." The sense of community is something that struck

Caroline as soon as she took up her role. "Looking out for each other is a way of life in Swaledale," says Caroline, who recognises that you don't need to be a churchgoer to have consideration for your neighbours. "Over the difficult winter months, the elderly are well provided for - they get their shopping done for them. We all look after each other. I do love the people and there is a really interesting mix; you have to have something about you, I think, to live here."

Caroline has responsibility for four churches: St Mary the Virgin at Muker which perches on a ledge of hillside in one of the loveliest spots in Swaledale; Holy Trinity, Melbecks, the church at Low Row, which belongs to the Small Pilgrim Places Network, a group of about thirty churches nationwide characterised by being "spiritual oases, somewhere to take stock and take breath"; St Andrew's at Grinton which has a history going back some nine hundred years; and Arkengarthdale's church, St Mary's, near Langthwaite, built

St Mary's Church near Langthwaite

Caroline at St Andrew's Church, Grinton

a few years after the Battle of Waterloo and one of around six hundred churches constructed at that time (so-called "Commissioners" churches) to encourage church-going during a period of dissatisfaction with the Establishment. "At all four churches there are lots of people who are willing to help," says Caroline. "They keep the buildings going and do the flower arranging and so on. There is a lot of invaluable work done by lots of people - even those who don't particularly come to services. A church is very much part of the community and a place's heritage. The joy here is that we can have every one of our buildings open every day."

No two weeks are ever the same for Caroline but her schedule generally includes ensuring that she conducts two morning services and one evening service in each church on a monthly basis. "Part of the interest of my job is that anything could happen - I don't really have a typical day.

There is a fair amount of admin to be done - with the four churches there are various committees to keep going. There is always a lot to prepare for Sundays: writing sermons and doing notice sheets and other more occasional things, such as banns of marriage." One of the most unusual aspects of Caroline's job recently has been her involvement in a bat project. "We have 'bats in the belfry' at Grinton church and we brought in a bat expert to take a look. The Yorkshire Dales National Parks Authority has given us some funding for a 'bat detector' and some professionally designed interpretation boards and we are going to celebrate our bats by having a launch of the project with a talk and a bat walk in September 2012. The congregation has got quite interested in it. It's introduced us to a whole aspect of the building that we hadn't thought about - as a bat habitat. It's great because it's another way of connecting with people."

The widespread population of Caroline's parish numbers around twelve hundred residents, many of whom she has visited at home. "I know a lot of the families," says Caroline. "You can just turn up uninvited in many homes and be made very welcome. However, this is Yorkshire and people have their ways of letting you know if a return visit is welcome or not," she laughs. "I have got to know the local families over the years through baptisms, weddings and funerals. There are now several families where I have done all three. In rural churches that is the core of it, really, and it's a big privilege to be part of people's lives at those important events." Recent changes in Church of England regulations allowing people to marry in a church where they have 'a qualifying connection' have meant that Caroline is now conducting more weddings than she was when she first arrived. "I like to personalise weddings using props," she says. "I once married a young woman who was a designer for Top Shop, so I had a tailor's dummy as part of the service and lots of fabric so she that she could 'design' a marriage. Another time I did an Everton-themed wedding for a groom who was a devoted supporter. We had the *Z Cars* theme tune which is sung by Everton fans and at one point I got out an Everton scarf and wore it for the rest of the ceremony. It makes things a bit more interesting and personal. I am very aware that a lot of the people who come to a wedding are not regular churchgoers, so you have to find a way of engaging them and to make it memorable for the bride and groom."

Since settling in Swaledale, Caroline has discovered that funerals are, in this part of the world, significant occasions especially amongst the farming community. "A Dales' funeral is a whole day affair - it's a cultural event like no other. The whole village turns out and it's not unusual to have several hundred people. There is a lot of food - the table is laid and re-laid. People come from a long way away and they all get fed. At a funeral my job really is to hold it together for everyone else. There are some funerals that get to you, but you have to do the professional thing - and sometimes that can be emotionally very exhausting."

Baptisms - of children and adults - in Caroline's churches generally take place during Sunday morning services as part of the main service so that the person being baptised can be welcomed into the Church family, but there have been exceptions: Caroline recently baptised five children in a river in Swaledale at the request of the Owen family, sheep farmers who live at Ravenseat, a remote part of the dale. "I know the family through their attendance at Christmas and Harvest Festival services at Muker church and through my connection with Gunnerside School where some of the children are pupils," says Caroline. "I was speaking to them at our Crib Service last Christmas and they were saying that their eldest had been baptised but that they hadn't got round to christening the other five children and they asked

Caroline conducting an Everton fan's wedding

me if we could do it in the river at their farm." Caroline was very happy to oblige because, as she points out, "Church is not about the building it is about the people inside - the Church family," so the whole congregation was invited to attend the event. "People were asked to bring a chair and a picnic," says Caroline. "It was a joyful occasion - farming friends and neighbours and the Church family from up and down the dale turned out in force. We shared lots of food, the sun shone and we sang *Glory to God in the Highest* as we gathered at the river." Caroline says she is keen to do more 'out of the church building' events and has previously held a Christmas Crib Service in a barn. "All the children dressed up as angels and we had a mother with her newly born child playing Mary and the baby Jesus. We also held a 'lambing service' on a farm as a Thanksgiving Festival."

Caroline first became aware that she had a calling to the Church when she was a teenager. "I came to faith at boarding school - we had a lady deaconess there and I remember talking to her about various questions I had," she says. "She encouraged me to get involved with a lot of things. When I was an undergraduate at Ripon and York St John, I was involved with the Christian Union and then worked with students before going to Durham for Theological College. It was a growing thing. When I was in my twenties I thought, 'When I am thirty I will do something about it...' so that was when I began to explore the possibility of ordained ministry. 2001 was quite a year - I got married three weeks before getting ordained!" Taking time to refocus on her own faith can be difficult to organise for Caroline with her extremely busy work schedule but she says that she has had to learn to make sure she has a balanced life. "I take Fridays off and nothing happens. Mostly, the phone doesn't ring and people give us a bit of space," she says. "I do need that time because doing this job does take it out of you. On Fridays Alex and I tend to go out of the dale or go and see family - it's a day for us just to be people. We sometimes go over to Saltburn and sit on the beach with a picnic. One day a month I go to the Jonas Centre near Redmire in Wensleydale which is a Christian retreat where I do some reading and thinking - it's a lovely peaceful space to reflect and reconnect with my own faith. People have certain expectations and everyone has an opinion about vicars and I think, in the past, some vicars have played up to a certain image. But I think you should just be who you are."

Tending to all her parishioners across such a large area means that Caroline has to be a good manager both of her own time and of the people who support her in her work. "Having to cover four churches, it is more of at team set-up, really. I have a few retired clergy who help out and they do as much or as little as they want to and we have a Reader for the parish. Readers can take parts of the service and I try to give them stand-alone projects: it's about playing to people's strengths. Readers are really important in rural parishes - they are licensed lay ministers, not ordained, but they have some training and they can do all the non-sacramental parts of the church service. In the Richmond deanery we share Readers so I can call on someone if I need holiday cover, for example: it's about using the resources available to you. It's a good team."

The *Vicar of Dibley* effect, as it is sometimes called, has been revolutionary since the first wave of women were ordained in 1994, the same year that the popular comedy series was first screened on television. More women are now appointed to the priesthood year on year than men, though many are serving in non-stipendiary posts - and, interestingly, many are taking on rural parishes. "In places like this, women are accepted," says Caroline. "We are keeping things going really, so I have never felt it to be an issue. It's only twenty years since women have been ordained but we have already made an impact, not least in that the Church has had to think about issues such as maternity leave. The Church is currently moving to a system called 'Common Tenure' which is a kind of employment contract for clergy, but we are not covered by employment law. It's the only job that can still be advertised where you can state that you don't want to employ a woman! Personally, I wish they had done it properly in the first place. The church has to move and grow and change - it's about life, keeping things alive and responding to people's needs." Caroline puts this philosophy into practice in various ways including conducting a service at St Andrew's on a Wednesday morning for those in her parish who prefer a traditional Book of Common Prayer service and she also holds a monthly service for around fifteen people in the sheltered accommodation in Reeth. "If people can't come to you, you go to them."

The interior of St Andrew's Church, Grinton

Caroline is keen to encourage inter-denominational relations, frequently arranging meetings with other church groups and organisations to co-ordinate Christian worship and fellowship. In addition, Caroline represents the Diocese at the Great Yorkshire Show as well as the parish at the Muker Show. She sits on the Swaledale Christian Council and on each of the individual church councils. She is involved in liaison work with the annual Swaledale Festival (which uses church buildings for its concerts), visiting sheltered accommodation for the elderly and compiling the parish newsletter - and she also has to update her Facebook page, a useful tool for connecting and communicating with the younger people in her parish. "I think a lot of people have an old-fashioned image of rural clergy as sitting around having holy thoughts and looking after your butterfly collection but actually there's always plenty to do."

Caroline also maintains links with the schools on her patch, visiting them and involving the children in church

Caroline enjoys driving up and down Arkengarthdale

events. "What is interesting is that the babies that I baptised when I first came here are now at the schools. Children from Reeth School have come to look around the church - I told them about the history of the building and spoke to them about baptism. At Arkengarthdale School I am a governor and as involved as they want me to be - I do assemblies for them and join in with school life." In 2011 Caroline oversaw a very special event involving local schoolchildren which took place in the orchard at Reeth. "We staged a Royal Wedding - we had a 'Will' and a 'Kate' and a best man and we talked about the wedding ceremony and the meaning behind it. The children designed a cake and we had 'fizzy champagne' made out of lemonade. I did a children's version of the wedding service - the children were all seven or eight years old and they had a great time."

Because of the demographic of the parish, there are not many young people who attend Caroline's church services - generally there is an older population and families with children. "Most of the young people move away to buy their first homes," says Caroline. "They do tend to come back to get married and have their babies baptised. I get to know mothers and children through the work I do with schools and we have children and family events at the churches for Christmas, Easter and summer. We have created a children's corner in Grinton church; we are always thinking of ways to make the churches welcoming and accessible to everyone. I like connecting with the families and the children. One year we made an Easter garden with the children at Grinton church and we told the Easter story; they really enjoyed it and we do mini holiday clubs. There was a Council-run youth group in Reeth but that has gone for now - the numbers go up and down."

After six years in post, Caroline feels settled and clearly loves her work, her parishioners and the place. "I love the landscape I work in. Coming over from Leyburn when you have Swaledale laid out in front of you - it is a magnificent panoramic view. I like driving up and down Arkengarthdale which is quite wild. We have a friend who worked in the Falklands for a while and he says that it reminds him a lot of there." The remoteness of some of the places she visits, and even living in a relatively large village such as Reeth, is markedly different from working in inner-city Leeds but Caroline has embraced the challenge. "It has been a steep learning curve. That's partly to do with the job - it's quite a big step up from being a curate to running a parish. Then there is the whole thing about working out how everyone is inter-related, being an 'incomer'. You have to learn to live with the place - you can't just nip out to the shops, Leyburn has the nearest cash machine and you can't get a signal on your mobile phone. You just learn to live differently."

Most of all Caroline appreciates the people, the sense of community and the variety of her work. "There's never a dull moment," she says. "There's just so much energy and commitment from local people. I do work hard to find out about the things that matter in Swaledale. For example, I have enjoyed learning about sheep since I have been here - you have to know enough to have a sensible conversation about them. I think they are really interesting creatures. This is a very beautiful place to be and it's a real privilege to work here. The best part of my job is seeing people growing in their faith. It is really satisfying when someone says, 'I have learnt a lot from your sermons.' When you are wondering, 'Am I making a difference?' it is great to get that sort of feedback."

Parish website:
www.swaledalearkengarthdaleparish.org.uk

5

*

Davinia Hinde B VetMed MRCVS

Vet, Wensleydale

"I always wanted to be a vet; in fact, I can't remember a time when I didn't want to be a vet," says Davinia Hinde who has been running Bainbridge Vets practice in the village of Bainbridge in Wensleydale with her husband Michael Woodhouse since December 2011.

"I enjoyed the James Herriot books as a child and I had a lot of pets but I didn't particularly have a 'Eureka!' moment, it was just something that I always wanted to do. Some people might think it strange - you spend so many nights on call, and being married to a vet doesn't help because if I'm not on call, then he's on call. It does take love and dedication." It's immediately clear upon meeting Davinia - who is energetic, enthusiastic and passionate about her work - that she has both in abundance.

Davinia met Michael in her first term at the Royal Veterinary College in London and, she says, they were both determined not to start looking for jobs until they were qualified. "And then we saw an advert in 2006 for two vets to work across two practices - one in Leyburn, with John Watkinson, and one in Bainbridge, with Adam Hurn," she says. "We explained that we weren't qualified yet and we came up and spent a day with both vets, going with them on their calls, and at the end of it they offered us the job. So when the time came, we qualified on the Wednesday and started on the following Monday! We took over from Adam last December when he retired but he can still do cover work for us - and it's great to have a locum who knows the practice and the clients really well."

Davinia and Michael live in the village of Carperby - co-incidentally where James Herriot and his wife Helen spent their honeymoon at the Wheatsheaf Inn - which is right in the middle of their extensive practice area that covers some

St Andrew's Church, Aysgarth, where Davinia and Michael married

of the most picturesque areas of the Dales. "It stretches right up to Ribblehead viaduct, across to Thwaite and up to Catterick Garrison, over to Grinton and right over to Kettlewell, so we cover a lot of Swaledale and Wensleydale as that's where the majority of our clients are. We do eighty-five percent large animals - most of the farming around here is dairy, beef and sheep. I tend to do a lot of sheep work as I am studying for my sheep certificate, which is based mostly on case work."

Inevitably in a practice that covers such a large geographical area, Davinia spends a fair amount of time on the road which can sometimes bring its own problems. "You have to learn which roads to go on and which not to go on, especially when there is a danger of flooding - which is quite often around here! And it can be quite hair-raising trying to

Davinia Hinde always wanted to be a vet

get about in the winter snow. I had to go over Kidstones Pass in the snow last year to attend a calf bed (that's when a cow has a prolapsed uterus) and that was a bit scary."

The couple do also hold morning and afternoon surgeries for what Davinia refers to as 'the smallies' - pets of the local community - which is an equally important part of the practice. Even at the busiest times, Davinia and Michael will ensure that one or other of them is available to do the surgeries. "Recently one of our clients who has a little Jack Russell terrier rang up on a bank holiday Monday to say that the dog was vomiting. So they brought him in and I gave him some treatment. Then he started collapsing and I wondered whether he had a rare condition that I had read about, so I took some blood and we sent the samples off to the lab." The results came back confirming Davinia's diagnosis and she was able to arrange appropriate medication from America. "He will be fine now, but if I hadn't spotted that then I would have had to put him to sleep." There are some instances when euthanasia is the only answer and Davinia is respectful of the strength of feeling that accompanies such an occasion when people have to make the decision to part with a much-loved pet. "It's an emotional experience for people and if we can do it in a dignified and sympathetic way, people really appreciate it."

Working in a rural practice, Davinia also has to treat a special category of agricultural animal - the working dog. She greatly admires the relationship that is built up between farmers and gamekeepers and their dogs. "There is so much training and effort involved," she says. "One of the worst things about my job is when I have to put a farmer's dog down. The farmers are distraught; they spend so much time with their dogs - because of the nature of their work some farmers end up spending more time with their dogs than with their wives."

The falls at Bainbridge

The Wheatsheaf, Carperby, where James and Mrs Herriot honeymooned

Lambing time, which takes place mainly in April across Wensleydale and Swaledale, is generally the busiest time of the year for Davinia when disturbed nights are the norm. "You just have to accept that you are not going to get much sleep," she says cheerfully. "Most of our clients are very sensible and they will only call us when they really need us but we do inevitably get a lot of calls in the middle of the night to do a sheep caesarean. Sometimes you have to do what you can in a bad situation: there are times when you just have to get a lamb out to save the sheep. I recently had to pull out a thirteen kilogram dead lamb from a sheep. In that sort of case it's battlefield surgery so, if you are presented with such a scenario, you have to do the best you can."

As a large animal vet, Davinia is dealing with people's livelihoods - some of the prize Swaledale sheep which are bred by many farmers in her practice area can sell for up to fifty-five thousand pounds. However, she says that she has never felt pressurised by any of her clients. "They all know that I always do my best and they trust me to do my job - it's important to have that relationship. The hours are long but, because our clients are so lovely, I can manage; I want to help them, so it makes it easier to do. I may not always be that cheery at two a.m. but it's easy because I want to make a difference for them."

Making a difference includes ensuring that flocks are healthy and helping farmers to take preventative measures. To that end Davinia and Michael run regular farmer training courses which have been welcomed by the local community. "The trouble with sheep, in particular, is that they try and die on a regular basis - there are so many things that can go wrong with them and they can be affected by so many different diseases. We give farmers vaccines which they can administer themselves - we do sometimes do vaccinations for smallholders if the farmer isn't confident about doing it or if they need a veterinary certificate - but generally the farmers get the vaccines from us to use on their farm. The sheep we deal with around here are mostly Swaledales, Dalesbred sheep, Blue Faced Leicesters and Texels. On the cattle side we have a lot of dairy farms with Holstein Fresian, Jerseys and some British blues and Limousins."

Davinia was still at college when the last outbreak of Foot and Mouth occurred and she admits that she would have found it difficult to deal with. "I don't really know how I would cope having to tell my clients that their animals would have to be slaughtered. You can still see the effects of the last outbreak. It takes a long time to get flocks and herds back up and to recoup the losses." Davinia believes that the next big problem to face the farming community will be TB and to safeguard against this she and Michael carry out rigorous testing programmes as required by DEFRA

(Department for Environment Food and Rural Affairs) throughout their practice. "We test all cattle for TB from calving onwards in a four-yearly testing area. If a cow fails the test, then it is taken off to slaughter. Our clients are vigilant and they buy locally wherever possible but with the way animal movements are, the risk is greater."

Davinia and Michael also carry out testing for BVD (Bovine Viral Diarrhoea) another infectious disease that affects cattle. Some of the most serious effects of the virus include abortion and infertility which can have a devastating impact on a herd and cause significant losses. "In Scotland they have a very effective eradication scheme for BVD which I think we should follow but it has to be industry and Government-led," says Davinia. "If the farmers aren't behind it, then it won't happen. If you are asking people to spend money on all the testing and then you cull animals without compensation, there is no incentive for the farmers to support the scheme."

Although Davinia is originally from Shevington in Lancashire ("Don't tell anyone," she jokes), she feels settled in the Dales and she and Michael are very happy living in Carperby where they have made friends and feel very much part of the community. "I think it is the people that make the Dales - the landscape wouldn't be the same without the farming families who maintain the land and the walls," she says. "Living here everyone knows each other and everyone is so friendly. When we first moved up here we lived in a practice house in Leyburn and then we settled on Carperby. As soon as we got there, everyone was really welcoming and they made sure we knew about events that were going on in the village; people want you to mix. Our clients have been very nice to us too. I think people like it when you are coming to live in the Dales to fulfil a purpose - they like people who give things back to the community."

A view of Bainbridge from the site of the Roman fort

There is no doubt that Davinia is giving something back to her local community - through her work, her thoughtful approach to her clients' needs and her no-nonsense, absolute commitment to getting the job done to the best of her ability. She has no illusions about what is expected of her. "You are only as good as your last case. People remember the disasters more clearly than they do the good things." She says that she has never experienced any sexism or been treated any differently because she is a woman. "Only in a nice positive way: I will get hot water, soap and a towel when I am attending a birth whereas it's cold water in a bucket for Michael," she laughs. "I don't think our clients care what sex you are - they just want you to prove yourself."

Making sure that her clients get the best possible service is a priority for Davinia and with this in mind, she has been working collaboratively with other vets in the North

Yorkshire area, through Dale and Vale Vets, a group of six veterinary practices set up in 2008. The tagline on their website is 'supporting agriculture in North Yorkshire.' "We set it up primarily as a farmer training system," she says. "We take it in turn to run the farmers' training sessions and we all have different and complementary specialist skills. We have monthly meetings; it's supportive not competitive - we don't poach each other's clients - and we help each other out. It's about local co-operation - sharing knowledge, experience, expertise and equipment, and exchanging ideas. It works really well." The seminars, workshops and farmer meetings that Davinia helps to run cover a wide range of issues in animal husbandry and healthcare and are held at a variety of venues - including farms. The events are well attended by vets and clients from all the six practices involved.

Many of the local farmers' children attend the training sessions that Davinia runs and she is impressed by their enthusiasm and desire to learn. "I love their keenness," she says. "When you are on a call, they are often around, watching what you are doing and asking questions. At the age of around twelve or thirteen they are already starting to build their own herd or flock. It's great to see that kind of interest and passion - and you need the younger generation coming through. I run a lambing course for youngsters and it's great fun being with them."

Passing on her expertise and knowledge is obviously something that Davinia finds very rewarding as she also regularly accommodates veterinary students for several weeks' training. "I really love having vet students at the practice - we always try and take some on each year," she says. "They stay for two or three weeks and the good thing from their point of view is that here they get experience of both farm and small animals. They can stay with us at our cottage and we take them on call with us. As long as they are keen, I'm happy and we get a lot who come back year after year - it's very satisfying to see how they develop and grow in confidence."

Despite a busy working schedule, Davinia and Michael do manage to have a social life in the village, although she says it can be a little unpredictable due to their professional commitments. "One or other of us is usually on call but we have a lot of friends. One of the many nice things about living in Carperby is that a lot of ages mix together. In some places in the Dales that doesn't happen because the younger people tend to move away and they don't come back again until they retire. There is a great community spirit - if there is a problem, people get together and do something about it. At Carperby Football Club recently, they needed some new changing rooms so everyone got together to do some fundraising." She belongs to the Ladies' Circle - there was a suggestion that she join the WI but she says that she will wait a while before she does that - and she is involved in a monthly book club. "There are four of us, all women. We take it in turns to choose a book and go to each other's houses for a meal and a discussion. A lot of our talk is gossip but we do also talk about the book."

Occassionally Davinia is able to have some time off and meet up with friends for coffee or to go shopping. However, she does sometimes surprise her companions with the contents of her handbag. "I will often have all sorts of odd things in there," she laughs. "Once while I was searching for my purse I pulled out a 'red devil' which you insert into a cow's stomach if they are bloated [they can't burp] and it lets the gas out slowly. You make a little incision and then insert it. They do come in handy." Clay pigeon shooting is another pastime that she enjoys which is popular with other residents of Carperby. "It's something that Mike and I can

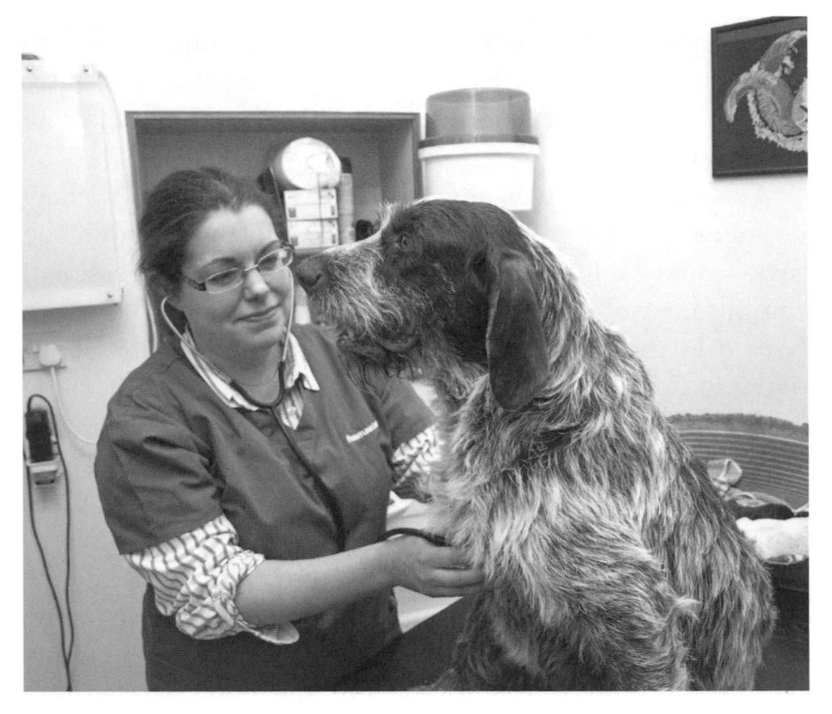

Davinia in her surgery

do together, although I go on ladies-only shoots sometimes too. My dad has always shot and Mike does a bit of game shooting as well - I don't. I have no problem with it, I just prefer not to. I have this joke that in Carperby when the siege comes the village would survive quite well," she laughs. "Everyone is a good shot and there is a range of people living there - we have farmers, doctors, dentists, teachers, electricians - all the skills that you would need to get by."

It is clear that Davinia and Michael, who got married at St Andrew's Church by Aysgarth Falls (one of her favourite places), have made a life that they are very contented with although working in rural practice is not a choice that suits everyone: according to Davinia there is a real shortage of vets coming through who are willing to take on the challenge. "It is difficult to get people for rural practices. If you work in a small animal practice in a town or city, you don't have the middle of the night call outs and, to be honest, I don't know if I would necessarily want to go to a remote area on my own - it is a lot easier as a couple. It is hard to find recruits."

There is no doubt in Davinia's mind, however, that the professional choice she has made is the right one for her. "I wouldn't want to be indoors all day. I could never see myself doing one hundred percent small animal work. There is a different kind of pressure doing large animal work because we are dealing with people's incomes, not just their pets, but I love getting my teeth into something and solving problems for them - doing investigations into infertility, for example. You can't recreate this situation anywhere else." The working relationship that Davinia has built up with her clients is one of mutual respect, with each understanding the other's commitment to providing the best care for their livestock. "We have a whole range of clients who we have got to know and are fond of," she says. "We are very happy and settled here - this is a job for life."

www.bainbridgevets.co.uk

6

*

Professor Dianna Bowles OBE

Academic and sheep-keeper, Upper Nidderdale

Dianna Bowles, who is originally from Leicester, has had a distinguished academic career. She has recently retired from her post as Chair of Biochemistry at the University of York where she has been teaching postgraduates and leading research groups since 1994. Her home, in the remote village of Middlesmoor in Upper Nidderdale, is an eighteenth century blacksmith's cottage, originally three dwellings, which she has renovated since moving there from Leeds in 1989.

Like everything she is involved in, Dianna brings passion, intelligence and enthusiasm to both her work and her 'pastime': keeping Herdwick sheep. She bought her first two Herdwicks in 1991 and now has around fifty breeding ewes but she is at pains to point out that she is a 'sheep-keeper', not a farmer, nor a shepherdess. She describes herself as a practical person, saying, "I don't see any point in being a daydreamer. I like making things happen."

How Dianna came to be living in the heights of Upper Nidderdale with relatively few neighbours derives from the perhaps paradoxical need she has for peace and quiet while wanting to be able to make as much noise as she likes without disturbing anyone. "I was at the University of Leeds and had lived in various places in Kirkstall and Headingley for many years from 1979 onwards. I was looking for a detached house because I need complete silence in which to work but I also like to listen to really loud music - and if you live in a semi that's a bit tricky. This cottage came up, I decided to have a look at it - and that was that. It wasn't so much the house but the drive from Pateley Bridge to Middlesmoor that just blew me away; it was magical."

The property which, when she first moved in Dianna says "felt unloved and quite forlorn", was originally a terrace comprising a forge, a butcher's shop and a shepherd's hut and it was only possible to access each separate compartment from outside. Gradually, Dianna opened it all up to become what is now a cosy and welcoming home upon which she has stamped her own distinctive mark. Her desk is by a window overlooking a wonderful view of the church and valley beyond. "I have done all my writing work at that desk for the past twenty-two years."

Commuting every day from Middlesmoor to Leeds meant that Dianna was able to feel that she was part of the village rather than a weekend-only resident - and the drive never felt like a chore. "I have often thought of the commute as like driving through various magical kingdoms," she says. "Sometimes on my drive back home from Leeds I would see moonlight on Gouthwaite Reservoir, which was simply amazing. Middlesmoor is situated on a hill and you don't see that very often in this country; it reminds me of a little Italian hill village." She quickly settled into village life

Dianna with some of her older Herdwick ewes

and still enjoys being part of the small community which numbers just over twenty. "Everyone is kind and friendly but I don't think you could live here beyond the age of about seventy-five - you have to be robust. Some people commute to work, like I did; some farm; some are retired; some have young children. Some are new and some have lived here all their lives. It's a good mix. What I love about the place is the silence, the beauty and the light. The light can be unbelievable - it shifts all the time. You get weather in capital letters in Middlesmoor. You can get thick mist with driving rain at the same time, and the snow blows horizontally here. I remember once when it snowed heavily in Middlesmoor and I phoned in to work to say I might have some trouble getting in to Leeds. I set off expecting the worst but was amazed to find there was no snow by the time I went down the hill to Lofthouse, the next village along the valley."

Being so high up in Middlesmoor makes a huge difference in the winter when the temperature can get down to minus twenty-six degrees, although Dianna says that in all the time she has lived in the village she has only been snowed in once, over the winter of 1990/91. "The roads aren't actually that bad up here because the farmers clear them. People who come to live in a place like this often have an impression of what it's going to be like. They come and they either stay for a year or they stay forever. For me, because I was going in to work every day, there wasn't really a period of readjustment." Her time spent commuting did, however, prove to have a life-changing effect. It was on her regular drive from Middlesmoor to Leeds University that Dianna became acquainted with a distinctive breed of sheep.

"There is a farm near Menwith Hill and the Sun Inn that has Herdwicks and I kept noticing that the fields in Spring would be full of these little black lambs, but not black sheep. In the end I stopped and spoke to the farmer, Harry Hinde, whose family are originally from Ennerdale in Cumbria. He had moved over to Yorkshire and got married and stayed and he'd brought over his Herdwicks with him." So fascinated had Dianna become with these sheep that she asked if she could buy a pair of little black lambs. Harry was happy to oblige since he was keen to build up a community of Herdwicks in Yorkshire, so he sold Dianna two shearlings and delivered them to her in Middlesmoor. She asked a local farmer if the lambs could join his sheep in a field nearby and Harry offered to come over and check on them from time to time.

Dianna can't exactly explain what it was that first prompted her to want to buy the lambs, although she admits that their attractive appearance was part of the appeal. "At that stage I thought I would put them in a field and I could just go and look at them," she says. "So they duly went into a field and I would go and look at them." In the meantime, Harry had put the shearlings to the ram: soon more lambs came along and Dianna found that she needed to find more space for them. "Some tenanted land came up, so I rented it. And so, not knowing a thing about it, I went into sheep-keeping," says Dianna. "It took me more than ten years to learn how to do it and I think I have only really learnt to see the sheep in the last five years, probably. It's the same with any animal - you can work alongside them without really noticing them or seeing them properly. When, eventually, you do, you can begin to appreciate them and know whether they are about to go down with something or predict where they will go. You, personally, move across a line, from outside to inside. If you are keeping sheep you really need to learn about them and their ways - all the youngsters in farming families I know around here, they go to sales and feed pet lambs from when they are children. It becomes part of their being."

Dianna acknowledges that her true trade for forty or more years has been plant biochemistry, not the rearing of sheep. "It's very strange coming to a completely new situation and having to learn again, but the Herdwicks are very good teachers. I have tremendous respect for the sheep and for their professionalism. I think that sheep are hugely underrated. I like them immensely and, watching their behaviour - sitting on a hillside as part of a flock - is wonderful."

In observing her sheep, Dianna has learnt a great deal, including the fact that sheep are only quiet if they are content. "The more they are quiet, the more you can interact with them. Usually it's mismanagement that causes problems - I don't agree with the phrase that 'every sheep is waiting to die'." Dianna believes that sheep are more intelligent than is appreciated. "It's been reported that they have the same neuronal capacity as dogs and they certainly understand simple phrases such 'go left' and 'go right' or 'come' or 'wait'." Dianna recognises each of her sheep as individuals. "They all look very different and have different characters!"

Once turned out on the moorland, they flourish. However, Herdwicks, she has discovered, because they are hill sheep and farmed in vast areas of open moorland, tend to have little concept of boundaries and fields. She relates how one ewe, brought from Cumbria, would simply climb straight over any wall she was confronted with. After a few weeks in Yorkshire, she did the same thing, leaping straight out of the pen, only to re-appear on the top of the wall to consider her options. "She chose to jump back in the pen and that sealed her fate to stay with me," says Dianna. That sheep is now twelve years old.

Dianna particularly admires the way her Herdwick ewes

A view of Middlesmoor

handle lambing. "They are so motherly. It's the Fell Ewe quality of them. You very rarely find one in the middle of a blizzard not knowing what to do. Unlike many other breeds, they cope well with triplets, bringing up three lambs without additional milk. I find all the ewes are really 'sorted'; every year I award a Mother of the Year prize and often that goes to a shearling yet, as first-time mothers, they have had no training at all. They can have their lamb dried, sucked and standing in an hour." Because Herdwicks are hill sheep, Dianna says she tends to 'lamb out' in the fields. "Bringing them inside for lambing just seems to put them off their stride. The lambs generally start arriving from the last week of March; we get a few every day and then all of a sudden there's a whoosh in mid-April when half the flock lambs in a few days. I have people who help me out at lambing time - it's pretty relentless."

Another feature of lambing time that Dianna noticed,

'The ancients', Dianna's older ewes

because of lambing 'out', is that when a ewe's waters break, she will stay where her waters have broken and remain there for up to three days afterwards. "The ewe keeps the lambs close to the birthing site and the familiar smell. Once the lambs are sufficiently bonded with their mother, they move away as a close-knit group. If you move the lamb away from the ewe prior to that, a first-time mother can easily lose her head and constantly returns to that site to 'find' her lambs. It has never really made sense to me, the agricultural practice of moving the ewe and lambs to a different field as soon as they are born." One thing Dianna has learnt from close observation is that, when permitted to live a full life, in almost all cases the relationship between mother and lamb lasts a lifetime. "The bond between the mother and the lamb is so strong."

Deaths at lambing time are an occupational hazard but thankfully rare in Dianna's flock. However, working with sheep has taught her how to deal with a grieving ewe. "If a lamb dies, often the carcase is moved away immediately. When I used to do that, the mournful bleating of the ewe was truly awful as she went round the field constantly calling for her lamb. In time I realised that, if I left the dead lamb with the ewe, within a day or so of remaining by its side she would cover it up and eventually return to grazing". When the young lambs are separated from their mothers in August, the calling from both the lambs and their mothers can go on for days and days. The lambs in particular will attempt almost anything, such as jumping cattle grids and walls, to get back to their mothers.

Observing sheep clearly matters to Dianna, but one can also learn from those who work alongside sheep. She tells a lovely story about being at a conference in the Swiss Alps and noticing a shepherd with his flock. "He had a portable electric fence that he would use to keep them together overnight. I watched him as he stayed with them. A lot of that involvement has been lost in how sheep are currently farmed. However," she adds, "I know it's very difficult to make an income from sheep without taking the numbers up." Dianna arrived at a point where she had to choose exactly what her future involvement with her sheep would be, particularly with regard to how many she could reasonably take responsibility for.

"Around five or six years ago I felt I had to decide why I was maintaining the flock. It is clear that I am not a farmer, nor are these animals pets. What I do is keep sheep, principally to observe them and work alongside them. I have fifty breeding ewes and then I have the little ones - lambs and shearlings; I usually take around thirty female lambs over the winter to select new breeding stock the following year. Most of the 'lads' are sold for meat at about six months old. It's not easy taking animals you work with

Dianna with some of her shearlings

Welcome to the Dales

to their death. I don't sell my animals at auction, I take them to the slaughterhouse myself and the meat is then sold through the butchers at Pateley Bridge. I eat lamb I have reared - the taste of Herdwick meat is superb."

Her commitment to her sheep is altogether apparent in the fact that she has now bought the land, within a mile or so of her home, where she keeps her Herdwicks. Unlike the vast majority of those who rear sheep, Dianna is not looking to 'turn a profit' so long as they cover their costs. Over the years she has increased the number of quality sheep, some of which she shows at the Yorkshire Show where she regularly wins prizes with her finer specimens. "Herdwicks are often referred to as a rare breed - but they are not, it's just that they are concentrated in the Lake District. Ninety-five percent of the breed is located around the Coniston area and there are tens of thousands living there. Their strong point is their hardiness - nothing can compete with the Herdwicks."

During the Foot and Mouth epidemic in 2001 Dianna established the Heritage Gene Bank to save bloodlines and breeds of sheep at risk from the outbreak, such as those breeds concentrated in areas where the disease was greatest. This work developed into the national charity, The Sheep Trust, which continues to work to protect the regional native breeds of the UK. She finds she is really interested in the genetics of sheep and has maintained extensive records of the pedigrees over the twenty or more years she has been breeding Herdwicks.

Dianna's academic work for four decades has been fulfilling, significant and beneficial. "I have always really enjoyed research. After a number of post-doctoral research posts in Europe, I started my academic career as a lecturer at Leeds University (from 1979 to 1993). I grew up academically there and progressed from lecturer to professor. Then I moved to the University of York, one of the reasons being that I thought I could commute there as easily as to Leeds. In fact, I found I couldn't, but fortunately I have always had some very good friends looking after the sheep during the week and I have always been able to take my annual holiday at lambing time."

One of Dianna's many initiatives at York was to create the Centre for Novel Agriculture Products. "I wanted to establish a centre where all the research and funding could benefit society," she says. "I think increasingly that scientists have a social responsibility to use their ideas and creativity to solve problems." One of Dianna's most recent projects has been funded by the Bill and Melinda Gates Foundation, investigating the development of a drug treatment (associated with the plant Artemisia) for malaria.

"The thing about research is that you don't ever stop thinking about it," says Dianna. "My head has always been in the world of molecules." Her major success back at Leeds had been the discovery of a novel means by which amino acids in proteins could be joined together. "I enjoyed that - it's very cerebral in a way, but it totally pre-occupies you for hours on end. I also enjoyed the research we did on how plants respond to injury. If a plant is injured the whole plant systemically responds - damage one leaf and signals are released that travel round the entire plant and increase the organism's readiness to withstand attack. Research never ends, but as a molecular scientist you don't actually ever see the molecule you are interested in, you can only use technology to follow the consequences of its existence and actions."

At York, her work took her in new directions to try to understand how a plant 'house-keeps' its affairs. Plants respond so rapidly to their environments that their cells are always changing - as she says, "to maximise opportunities and minimise risks." Change can be due to the weather, an attack by a pest or pathogen above or below ground, drought or flooding and so on. Her research followed the activities of a family of enzymes deeply involved in this house-keeping: she referred to them as the plant "calling in the A-team." And, interestingly, she has found the same enzyme family is also involved in making many of the plant products we use as medicines and dietary supplements or flavours and fragrances.

One of Dianna's many passions is to encourage interest in science. To this end, she successfully set up a project taking microscopes into primary schools to introduce young children to the wonders of biology. "I think education is about inducing curiosity. When five year olds look down microscopes they see all these things they don't normally see and then they get curious. Biologists can have difficulty in explaining things to very young children, so I got Yorkshire Arts involved - we had artists and scientists working together in the primary school classrooms and the children painted what they could see." Working with the children brought back memories of Dianna's own childhood and how she had been inspired herself by looking down a microscope. "I can remember borrowing one from school and looking at a puddle of dirty water - there was a whole new world in there! After that I only ever wanted to do Biology, understanding why things did what they did and, forty and more years later, I still do!" Preparing for retirement from her academic work which has been such a big part of her life has, she says, been strange, but she has plenty of ideas about how she might spend her time. "I want to carry on researching by writing reviews as well as science writing for general audiences. I particularly like outreach because I love communicating about science."

When she is not thinking about science or looking after her sheep, Dianna enjoys a variety of different

Herdwick shearlings

entertainments. "I never get bored. I adore film and going to the cinema but from Middlesmoor the nearest cinema is in Harrogate twenty-three miles away so I tend to watch films at home on DVD. I really enjoy photography too, but I think what I like most is being outside." This love of the countryside, and an awareness of the joy it brings, has developed into an interest in providing quiet places for people who would benefit from an escape from the pressures of their lives - and she has recently set up a charity, Aelred's Grange, to do just that. "I'd like to help carers, and the cared for, to enjoy the beauty of this countryside." She is starting to maintain a flock of older ewes that she refers to as 'the ancients'. They no longer lamb, and live out their natural lives in the Nidderdale fields. These sheep, Dianna feels, would be an ideal group for visitors to interact with. "They are very calm and calming," she says. With all the pressures associated with her competitive academic career, it is very apparent that the Herdwicks have been useful 'stress-busters', as Bridget Kendall described them in a recent Radio 4 *One to One* interview with Dianna.

Around the land, which Dianna describes as a place of

Alison O'Neill with her dog Joss at Shacklabank Farm

Gamekeeper Amy Lucas in Cotter Dale

Annabelle Bradley at her smithy in Malham

The Reverend Caroline Hewlett at St Andrew's Church in Grinton

Vet Davinia Hinde at her surgery in Bainbridge

Dianna Bowles with some of her Herdwick sheep in Nidderdale

Gillian Howells, Creative Consultant, in Richmond

Curator Helen Bainbridge outside her museum in Reeth

Ethical entrepreneur Isobel Davies with some of her Shetland sheep in Richmond

Kimberley Brereton, landlady of the Foresters Arms in Coverdale, *below*

Artist Moira Metcalfe at work in her studio in Appersett

Pat Thynne in her renovated farmhouse in Grisedale

Lettercarver Pip Hall in her studio in Cowgill

THWAITE woolcomber, & Isabella
1817-35 Henry LICKBARROW cor
wainer, & Margaret 1823-
Mathew EDMONDSON tailor, &
Betty c.1829-81 Thomas ALL
farmer & stonemason 1871-c
William ALLEN coal agent, & Mart
c.1891-c.1901 Simon PARKER pla
layer, & Martha 1901-12 Edwa
OVERSBY platelayer, & Elizabe
1911-c.18 James NELS
stonemason, & Agnes c.1918-
Ned & Agnes MIDDLETON
1951-72 Jack & Kate AKRIGG
1972- David & Anthea BOULTON

Zarina Belk in the kitchen of her tea room in Kettlewell

"utter beauty", there will be benches for people to sit and enjoy the landscape. There are also two barns on the land which Dianna has recently had renovated. "One of them had three trees growing in the middle of it!" she says. "These barns date back to the mid-seventeen-hundreds and one in particular was so beautiful and so derelict, I felt I had a responsibility to recover its existence."

Dianna converted to Catholicism when she was eighteen but by thirty had lapsed. Four years ago, she started going back to Mass at St Aelred's, a Catholic church in York, influenced by the death of her mother at the age of ninety-six. "I spent the last forty-eight hours of her life with her. We had reached a wall, a place of being in which I was at a complete loss as to what to do, because there was nothing I could do. Much to my surprise I found myself saying Hail Marys. After her death, I thought that if it had seemed so right to pray, then I had better consider the possibility of returning to the Church. I think we are always so full of busy-ness and thinking we can make a difference that there is no space or place for us to recognise the presence of Our Lord. I really don't know how strong my faith is, but I believe in joy and kindness and caring, so that's a start."

Dianna's life has been driven by curiosity, competitiveness in her career and a pervading sense of joy. As she has got older, she appreciates far more how her sheep-keeping has helped her philosophically and spiritually. "Being outside in the natural world brings an immense sense of quietude - it anchors you in reality."

www.herdwicksheep.co.uk

www.aelredsgrange.org

7

*

Gillian Howells

Creative Consultant, Swaledale

After thirty years of living and working in Richmond, Gillian Howells has become one of the town's most influential figures.

Even though she is now in her early sixties, she has the energy (and commitments) of someone much younger. Indeed, in listing the range of her activities, it is clear that she undertakes what would take three people or more to replace her. She describes herself as a creative consultant, with one of her roles being to promote and sustain North Country Theatre, the much loved itinerant small-scale touring theatre company founded by her husband, Nobby Dimon, who is also the theatre's artistic director. She is the secretary of the Richmond Business and Tourism Association (RBTA), the first female Pasture Master of the Richmond Burgage Pastures Committee, founder of the Arts in Richmondshire (AiR) organisation, co-founder of the now well-established annual Richmond Walking and Book festival, a mentor for Dru Yoga classes and a support to a number of local organisations who require her business and marketing skills in developing their enterprises.

Gillian's love of Yorkshire, and Richmond in particular, is very apparent but she is originally from rural Shropshire where her father was a pharmacist. Her interest in drama forms part of her earliest memories: "I think I had a talent for acting when I was younger - there are lots of photos of me performing. My mother was completely deaf in one ear so, whenever I performed, I projected my voice so my mother could hear at the back of the hall. I went to drama classes on Saturdays in Shrewsbury and my Saturday job as a teenager was helping out with the children's classes. I knew that's what I wanted to do and I got onto a course at Dartington College of Arts studying drama and dance. The third year there was for teacher training. I worked for a theatre company initially then moved into teaching at a girls grammar school in Manchester and then in a comprehensive

Gillian in her office in Richmond

in the East End of London. There was no grass, which was a bit of a culture shock!"

While serving in Germany for sixteen months as a youth worker for the British Forces, Gillian became involved with British Forces Broadcasting. On her return to England in 1977, she found employment at the Dovecot Arts Centre in Stockton on Tees assisting with community arts projects, working on local estates, offering drama workshops and engaging in development work through the arts. "The local radio station, Radio Teesside, used to use us for character voices on commercials and eventually I went to work there in 1978." She had been recruited by Donald Cline, the Commercial Production Manager at the station, who also owned and ran a bookshop in Richmond with his wife. He was subsequently one of the driving forces behind the Richmond Station development, converting the old railway terminus into a cinema, arts centre, restaurant, gallery and niche shopping centre. Gillian was involved in the project in its early stages in an advisory capacity, devising their audience development plan and putting Donald in touch with the community group who had raised funds for a similar restoration at Farfield Mill in Sedbergh.

Gillian's first visit to Yorkshire had been for the wedding of a college friend who was getting married in the Quaker Meeting House in the village of Bainbridge in Wensleydale. "That's when I discovered the Dales," says Gillian. "And I loved them! When I was working in radio in Stockton on Tees, Nobby and I used to come out here to walk and enjoy the countryside and then, when I got a job in TV, we moved here in 1982." The last thirty years have been testimony to just how much they fell in love with Richmond.

Gillian's work for radio subsequently led her into a significant period in her portfolio career when she worked in television. "Initially, I was with Yorkshire Television doing on-air promotions and then went freelance, working for a lot of the regional ITV companies like Grampian, HTV, TSW and Thames TV. I did a lot of work at YTV and joined the staff there, but I left in the first tranche of major redundancies and took the opportunity to set up my own independent production company. By that time Nobby and I had been in Richmond for about ten years and it was really nice to be here during the week rather than commuting."

Gillian continued in freelance production and writing, mostly creating links for Children's ITV. "I was on a two week cycle and worked with such 'stars' as Roland Rat, Christopher Biggins and the Krankies. I was also involved with video production." However, at a time when lots of other independent companies were forming, Gillian had to respond to the fact that the amount of work becoming available was more thinly spread. "I had to rethink. One of the functions of a producer is to bring people together and make things happen so, using those skills, I gradually moved over into creative and business development," says Gillian, and that is what she has been doing ever since, in amongst an array of voluntary activities.

Meeting Nobby was, very obviously, a hugely significant event in Gillian's life and it was through her initiative that the relationship blossomed as it did. "Back in the late Seventies I bought a house in Stockton but my social life was based around the arts centre - it was a really buzzy place with lots of activities going on. Nobby, who is originally from Doncaster, went to work there in 1979 - he had been working in a comprehensive school in Dorset and he decided to leave after three years. I met him at the Dovecot but he also did some voice work at the radio station and I asked him out." They have now been married for thirty years and still laugh together. "He says he married me because I am his best audience - that's not true because he's

A view of Richmond Castle

a very funny man and lots of people think he is amusing." Gillian and Nobby have always been able to arrange it that when one of them has been working freelance then the other one has had "a proper job". Nobby's background as a teacher as well as an actor led to his being Theatre in Education Director at Harrogate Theatre for eight years where, says Gillian, he had great success going into schools doing week-long projects with primary school children.

"When we set up North Country Theatre in 1995," says Gillian, who is a member of the company's board of trustees, "we got a Yorkshire Arts grant and did our first tour the following year with *The Thirty Nine Steps*, a version of which is still running in the West End. It won an Olivier Award and was a hit on Broadway - but it started here in Richmond! We also got a Scottish Arts Council Grant and went on a tour of

Gillian in Richmond

the Highlands and Islands, which was very appropriate for *The Thirty-Nine Steps* as a lot of the action takes place in Scotland." Though Nobby had a major involvement in devising the show, his current share of the royalties amounts only to a half of one percent. "He's very sanguine about it," Gillian observes.

Despite the continuing popularity of North Country Theatre, which performs in non-theatre spaces in rural locations from the Borders to Shropshire as well as at the Georgian Theatre in Richmond, the company recently lost its regular funding when Arts Council England reduced its funding portfolio. Gillian's response was typically phlegmatic and pragmatic. "We launched a Be An Angel of the North Country campaign and we now have twenty-seven thousand pounds a year from our Angels committed for the next three years. We have had an anonymous donation of nine and a half thousand pounds and the membership of our Friends Scheme has increased by fifty percent - it costs twenty-five pounds per household to join. All this has helped us bridge the gap left by losing our regular funding. We are now having to take the approach that, if we don't get a specific grant for a project, we can't do it. However, we have got European funding for a series of workshops for a Dales-based community play for 2013 - that's very exciting - based on two pieces of local folklore, The Reeth Bartle Fair and the Burning of Bartle from West Witton. It's a sort of *Per Gynt* for the Dales. We hope to make a link with our Norwegian twin district, Nord Fron Kommune, which is co-incidentally the home of the *Per Gynt* legend."

The last time North Country did a big community play was in 2005 with *Last Dance of a Dalesman,* which brought together people of different ages, abilities and interests from around Wensleydale and Swaledale. "We performed it in various venues across the district," says Gillian. "It gave a large number of people the opportunity to work with professional actors and a professional director. The story was about the last lead miner in Yorkshire combined with legends relating to the Corpse Way, the funeral path through Swaledale leading to Grinton." Gillian hasn't given up thoughts of going back into acting herself: she says she has ambitions to stage a one-woman-show about authors Marie Hartley and Joan Ingleby's Yorkshire Dales.

As well as running her office in Richmond in the building - part of what was once a Georgian coaching inn - where North Country Theatre is based, Gillian has applied her skills to assisting other local organisations, particularly with regard to bid applications for grants. "I have done big funding applications like Hudson House in Reeth - a rural enterprise and resource centre for Swaledale and Arkengarthdale - and then smaller things like the Red Cross charity shop in Richmond."

Hudson House opened in 2003 and enables local people and visitors to the two dales to have access to a range of facilities which otherwise would only be available in towns. Its popularity meant extending into The Barn, located nearby, with full disabled access. Various organisations pay a fee to operate within the premises, permitting Hudson House to be financially autonomous: such groups include the North Yorkshire Police, The Yorkshire Dales National Park Authority, Richmondshire District Council, North Yorkshire County Council, the Swaledale Festival, Richmondshire Citizens Advice Bureau and the District Community Transport project as well as business link groups for the area. Gillian's involvement was a significant contribution to the enterprise's viability.

Becoming involved with the Richmond Burgage Pastures committee has also led to the distinction for Gillian of becoming the first ever female Pasture Master. Hundreds of years ago, if you possessed a 'Burgage Right' then you were entitled as the citizen of a town or city to graze one of your animals (often a horse) on designated land at limited times of year, generally between April and October. In Richmond, following an Act of Parliament in 1853, holders of these rights have been able to elect a committee to manage the Burgage Pastures for them. The Pastures also include the former Richmond Racecourse which was registered as Common Land forty years ago and was subsequently included within the Countryside and Rights of Way Act (2000), giving the public the right to roam. The duty of the Richmond Burgage Pastures Committee, presided over by Lord Ronaldshay and supported by Gillian, is to ensure the maintenance and protection of the racecourse, its buildings and surrounding land. The grandstand, built in 1776 and designed by John Carr, is of particular historic interest and is Grade II listed. The racecourse is considered one of the finest early courses anywhere in the country and the grandstand worthy of restoration so that it can be removed from the English Heritage register of buildings at risk. The course officially closed in 1891 but the land has been used for training racehorses ever since. "From the racecourse you can see right out to Teeside, Wensleydale and the Vale of York - it's amazing," says Gillian. "We are working with the Architectural Heritage Fund and the Richmondshire Buildings Preservation Trust to get enough funding both to preserve the grandstand and get it off the 'at risk' register. It's a key time of transition." Gillian's extensive experience in fundraising and community engagement will be of great value to the project.

Another involvement is with the Richmond Business and Tourism Association which works for the benefit of members, their businesses and the wider community. "We meet once a month," says Gillian who is secretary of the association. "It is a member-led organisation helping to keep connections between various businesses in the Richmond area. We got some Lottery Funding and we ran a conference in March this year. Richmond is a Fair Trade Town and I'm also involved with that. The idea is to raise the profile of Fair Trade and to get the message across that how you shop affects the choices you make." One thing always seems to lead to another with Gillian, so it is little surprise with her theatre role that she became involved with the arts more generally in her local area. "I set up Arts in Richmondshire (AiR) which got a lot of things going. We put in place both an arts and a cultural strategy, we got funding for an arts officer and we had somebody working across the whole district. I've been able to meet a lot of amazing creative people in North Yorkshire - we are blessed with some really talented artists and makers. I was involved in the wider arts community helping to write the grant applications to support the creative industries in North Yorkshire and that has provided a really good legacy. I also set

up the websites for the Voluntary Arts sector." It is impossible to disagree when Gillian adds, "I think my strength is looking at and identifying solutions and setting things up, problem solving and making things happen. I like bringing people together. I am a background rather than a foreground person."

Gillian was very influential in the inauguration of the Richmond Walking and Book Festival. "I was asked to set it up back in 2004 and it has been running since 2006. I said, 'You have a lot of good walking around here, the last independent boot maker in the country and one of the best independent bookshops in the country. Why don't you combine them?' That was its Unique Selling Point. 'Boots and books' is quite unusual and now people come up specially. It has developed and is still going strong. Unfortunately, I had to back out of organising two years ago because of health reasons but it has kept going without me."

Now fully recovered, Gillian's well-being and obvious energy must partly derive from yet another of her activities -

teaching Dru Yoga and meditation classes. "I started doing yoga in Richmond in the early 1990s, trained to be an instructor and then set up classes. In the last three years I have undertaken further training so I can now offer Dru meditation. It's all about relaxation and getting the physical hindrances out of the way. It is very satisfying and this last month someone said to me, 'You have transformed my life with yoga.' I do quite a lot of classes in Richmond and also up at Catterick Garrison where the numbers can be fluid because people are sometimes away on manoeuvres or overseas duty. The principle of Dru is 'all that you need you have within you'. I have had a couple of people who have gone on to do the training themselves. Having taught for ten years my message remains that you use yoga and meditation in everyday life. It's funny how people don't make space for themselves. I think people who come to my classes are looking to keep fit and healthy but also to find that quietness within. The benefits are all round. How people perceive themselves, if you are proactive about it and take responsibility for your own health, giving yourself time to be quiet - that's all to the good."

Gillian's fascination with her adopted county shines through when she describes her favourite parts of the local area. "The top of Kisdon is a classic view, isn't it? One of the walks Nobby and I regularly do is to Marske. At the top end of the valley you feel as though you have arrived at Cold Comfort Farm. There are huge stone gateposts, and peacocks wandering around. That's lovely. Ivelet bridge and the walk all the way along to the Muker meadows is just wonderful. We like to repeat walks through the year so that we see places and how they change with the seasons. I have done all sorts of work - some of it paid and some of it not paid - but I remember having to drive to various places in the Dales and thinking, 'I'm being paid for this!' Bishopdale is particularly lovely. The landscape around here gets into your soul."

In 2003 Gillian and Nobby signed up to do a four-day labyrinth-making workshop with land artist Jim Buchanan at the Yorkshire Sculpture Park near Wakefield. With their new-found skills Gillian and Nobby created a large-scale labyrinth that people could walk around for the 2006 Walking and Book Festival. "It was fabulous," says Gillian. "We made a large eleven ring grass-cut labyrinth based on one at Chartres Cathedral. It was in a lovely setting by the river below Culloden Tower and, with the sun shining through a huge beech tree, it was like having a natural stained glass window. And we've been making them ever since. North Country Theatre made a huge one at Rievaulx Abbey. For me it brings together the contemplative with the land and nature."

For a year after moving to Richmond, Gillian commuted to London and would take the early train and last train back so she didn't have to stay away overnight too often. "Richmond is right in the middle of Britain, so it's more accessible than you might think - you can be in Oxford Street in three hours - but when I'd get back late at night, walking through the old part of town, I can remember the physical sensation of feeling, 'I'm home.' In Richmond it's not just the landscape, it's the built heritage too. I love living here - there is a real sense of community. It can sometimes take me an hour to walk to work because I keep stopping to chat. It's a fabulous place to live and we have great friends. In London it's really strange that people don't make eye contact at all and I think it's because space is precious to them and they are guarding their little bit of space. One of the delights of Swaledale is that you can get away completely and not see anyone. Richmond is just big enough to retain some anonymity. I had to get London out of my system but this is where my soul is comfortable."

www.creativenorthyorkshire.com

8

*

Helen Bainbridge

Museum curator, Swaledale

Hidden away behind the green in the village of Reeth in Swaledale is one of the small cultural gems of the Yorkshire Dales - the Swaledale Museum, run with obvious affection and attention to detail by Helen Bainbridge (aka Dr Helen Clifford) and her husband Alan. A wonderful treasure trove of a place, packed full of artefacts relating to the history of Swaledale and its neighbouring dale Arkengarthdale, the museum was originally set up in 1973 by a local couple, Donald and Erica Law, and Helen and Alan took it on in 2004.

"We had a holiday cottage in Reeth and we used to come up and spend every summer here," says Helen. "Since the cottage, which is now our home, is next door to the museum, we got to know Erica Law quite well and we used to help her out when we were here on holiday." Erica had been running the museum on her own since the death of her husband and mentioned to Helen one summer that she would like to retire. "She said that her family weren't interested in taking the museum on and she asked us if we would like to take it over. We thought about it and decided, 'Why not?'" The museum is a private business and receives no external funding so taking it on was a significant commitment both financially and professionally. At the time Helen was working for the Victoria and Albert Museum in London teaching on an MA course in the History of Design department run with the Royal College of Art, while Alan, a potter by training, was teaching art at the City of London School for Boys. "It was quite a big decision as I had had a long involvement with the V&A on and off since the early Eighties when I was a student there," says Helen. "But Alan and I were both ready to do something different and running our own museum really appealed. I think part of the reason was that, although I am an academic, I am also

Helen in the museum's cottage garden

quite practical - and if you have a small museum, you have to do everything."

The museum is housed in the old Methodist School Room, originally built in 1836 on the site of two late seventeenth/early eighteenth century cottages and gardens. In 1862 the building became a Sunday School after a new Quaker School was built in the village. "During the Second World War, the building was used to billet troops attending the six-week Battle Training course at Catterick Garrison," says Helen. "We sometimes get elderly gentlemen coming in who remember staying here as young soldiers and during restoration work we found tins of boot polish and packets of Woodbine cigarettes under the floorboards dating back to that time." After the war, the building was used for a variety of recreational purposes, including as a badminton court.

After Helen and Alan bought the building, they put all

A view from the village green at Reeth

the objects in trust to protect them for the future. "One of the first things we did when we got here was to set up a Friends of the Museum Group - and that was key because we have built on that. There are now around two hundred Friends and they are very good at turning out for events that we put on. We have really good connections and a lot of local support." The museum, which is open every day except Saturday from May to September each year, is run with the help of volunteers. There is a bank of about twenty with a core of seven or eight regulars and many of the volunteers are retired people who have moved to the area. "The number of people who live here to work is getting smaller and smaller," says Helen. "But it does mean that we get really interesting people helping us out and they are all full of enthusiasm. We have retired deans of universities and professors of geology - a huge skill base - and we are very fortunate that we have that kind of expertise to draw upon."

There are also, occasionally, younger people helping out since Helen established links with Durham University. "I was invited to give a talk at a Women's Institute group and I got talking to a lady afterwards who turned out to be a professor of Archaeology at Durham University. I talked to her about how we could work with her students which led to one of her PhD students coming to help at the museum creating a database of our objects." Helen admits that she had certain preconceptions about the WI and had not expected that outcome from her visit. "It just shows you that you should never assume things about people," she laughs. "It's important to be open to all sorts of connections."

Making those links is clearly something that Helen is very good at - she is bubbly, enthusiastic, approachable, passionate about what she does and very much a people person. "I am constantly looking at ways in which we can use the space and engage with people. The museum is a

Some exhibits at the museum

very lively place. It's a living place, not a dead place. A lot of visitors to the museum say how lovely it is to be able to touch things - it's about the magic of objects." Helen is the cheerful and welcoming hub of a variety of satellite groups that have formed as a consequence of her desire to share local historical knowledge, create a comprehensive and accessible archive and to enrich the quality of exhibits displayed. "I really like the fact that people trust us with their precious artefacts," says Helen. "They know that we will look after them. New things are brought in nearly every week and we have a substantial archive now of diaries and papers which people can come and look at here at the museum which will soon be accessible online. We get a lot of people coming to research their family history." Since she has been at the museum Helen has encouraged the setting up of a number of groups and projects including the

Vernacular Buildings Study Group, the Poor Law Project, the Swaledale and Arkengarthdale Archaeology Group, a craft group, a knitting group and the Swaledale Voices Oral History Recording. The museum's archive, which Helen initiated, is a valuable resource for researchers, academics, local historians and genealogists, and is connected to the North Yorkshire County Record office, thanks to the enthusiasm of manager Keith Sweetmore. With typical attention to detail, Helen and Alan have even restored what was the schoolyard at the front of the museum into a cottage garden, planting the type of vegetables, fruit trees, herbs and flowers which might have been grown before 1836.

The local drama group, which was disbanded many years previously, re-formed partly thanks to Helen. In 2006, amongst some boxes that were about to be thrown out after a house clearance, she found a 1940s pantomime script that had been performed in the museum by the Wesley Guild in 1949. The panto was a version of *Beauty and the Beast* written by a local woman called Mary Hilary. Jenny Curtis, the treasurer of the Museum Friends Group, suggested that the community might like to perform it that Christmas. "I never thought we would actually do it," says Helen. "But the response from the community was absolutely amazing." The script was updated and performed, with some of the original actors amongst the cast, amidst considerable media attention - including a spot on the long-running Yorkshire Television series *Dales Diary* - and the drama group is still going strong. "In fact, they helped out on a fundraiser I organised recently," says Helen. "I have boxes and boxes of historic underwear - I can't display all of it - and the drama group did a whole show with underwear as the theme. They raised a thousand pounds which paid for some essential building work we did over the winter."

Maintaining the building is just one of the many

responsibilities involved in running the museum as well as continually looking for new ways in which to attract and engage visitors - and Helen does this in conjunction with other academic and curatorial work. "It's a precarious life but it keeps you on your toes. It's like one mad experiment! We are living how we want to live, but it annoys me when people say, 'You are living the dream'. It is a dream but we are taking a risk. We run the museum but we also have to do work outside to keep the museum going. We have renovated the cottage next to the museum which we let out as holiday accommodation. The income from that helps pay for repairs and maintenance of the museum building. The idea is that if one fails you have something else to fall back on. You never know what's going to happen from year to year. There are no guarantees - every year is a fresh challenge - but you only need one person to come into the museum and say how much they have enjoyed it and it really buoys you up."

After graduating from Cambridge with a degree in History, Helen worked as a volunteer in various museums and as an archivist at the university's Trinity College. "I realised I was never going to make a good archivist because I just kept reading the documents instead of logging them!" She then went to the Royal College of Art, where she met Alan, to study for an MA in History of Design, that is, studying History through objects, which was run in conjunction with the V&A. "I really enjoyed the course and then I ended up teaching there. I loved the fact that we were looking at objects as source - they have multiple stories." Following her MA, Helen began a PhD which related to the archive belonging to Garrard, the Crown Jewellers from the time they were appointed by Queen Victoria in 1843. The V&A had rescued the archives, containing papers dating back to 1735, which would otherwise have been destroyed. "There were customer ledgers and all sorts of records," says

Helen in the museum

Helen. "I just fell in love with them! And out of that came my PhD. It was a socio-economic history, linking the archives to the objects."

Helen then went to work in the Department of Art History and Theory at Essex University where she taught on the MA course in Gallery Studies. While at Essex, she received an offer from the Ashmolean Museum in Oxford, a department of the University, to become a Leverhulme Research Fellow looking at the possession, caring for and keeping of the silver collection by Oxford colleges. "In 2004 I organised a big exhibition - *Six Hundred Years of Oxford College Silver* - and wrote a book in conjunction with that. So, 2004 was quite a momentous year - I published the book of my PhD, curated a major exhibition with accompanying publication and we moved up here!"

Helen continues to pursue her academic and curatorial work alongside running the museum in Reeth and her latest exhibition - *Gold: Power and Allure 4,500 years of gold treasures from across Britain* - opened at Goldsmiths' Hall in London in June 2012. She is also employed by the History department at Warwick University and is currently working on two projects connected with the East India Company as a senior research fellow and as Museum Consultant.

"It wouldn't work full-time just running the museum," she says. "There is the constant question of how you make it survive. But it is very good for me to be so grounded; you get a real insight into how a museum operates and an appreciation of all the different facets of an institution, such as marketing, for example." Finding new and different ways of appealing to visitors is an area that Helen has developed in the relatively short time that she and Alan have been running the museum. "We organise study days with experts in various different areas and we have a series of lectures and workshops. We do a whole range - all sorts of weird and wonderful things. We have had felt-making, paper-cutting and knitting workshops and we did a day in honour of Lawrence Barker who is now in his eighties and was the first ever National Park warden. He is well known and liked in this area and the event was packed out. We can experiment with the space."

One way of using the space differently is through visiting exhibitions which Helen initiated very soon after taking over the museum in 2004. "Our first was an exhibition of the work of potter Monica Young. She was French but lived in Reeth for many years and she made these huge five-foot pots. She had recently died and we got the opportunity to show her work before it went off to bigger collections elsewhere," says Helen. "I really enjoy the visiting exhibitions and giving people the opportunity to use the space. Being a local museum doesn't mean it has to be second rate. We are trying to work to a higher standard."

More recently Helen has exhibited the work of photographer Stuart Howat and artist Tessa Asquith-Lamb. Their exhibitions both highlighted the importance of objects in people's lives - very much in keeping with Helen's philosophy. "Their work is wonderful," says Helen. "And the shows reflected how we all create museums of our own through our treasured possessions." Tessa's etchings and paper-cuts were, in part, a response to the museum collection and included delicate bridal shoes, letters and fragile valentines while Stuart invited people to come to the museum, bringing one of their favourite things with them. He then photographed them outside the museum in natural light with their special objects. The resulting black and white collection shows people of all ages holding a variety of items including a silver box, a boat tiller, a food mixer, a skateboard and a tray of Lego.

Of all the exhibits you can see for yourself at the

museum, which amount to around two thousand, Helen's favourite is a sampler, dated 1862, made by Elizabeth Thompson aged thirteen. "It is special to me because it was one of the first objects that I tried to find out more about and also because it relates very specifically to the time when the building was changing from a day school to a Sunday school. I took it around to local Women's Institute groups and asked people if they knew anything about it. Eventually I met a lady who was related to Elizabeth Thompson, the girl who had stitched the sampler on the occasion of the re-opening of the Friends' Day School in Reeth, which meant the building which is now the Museum lost its day pupils to the more modern, smarter neo-Gothic Quaker School. It's a lovely example of how you can discover things by talking to people, listening and waiting."

Helen with the sampler which is her favourite object in the museum

Running the museum and being the creative driver for its further development as well as holding down a high-profile academic career leaves Helen little time for complete relaxation but, with three much-loved dogs in the household, she does at least have the excuse to go for a walk every day. "It's lovely to get out into the open air, even if it's raining - it's just good to get out," she says. "It's what you don't get in the city. Here you can just walk out of the back door and within minutes you are in glorious countryside. It makes everything better. I am an early morning person - we are often up at five a.m. and take the dogs out for a walk. My idea of a good walk is not meeting anybody for the whole time I am out - that's brilliant. The landscape is so important - if you regularly walk in it, you see it changing with the seasons, you get to know it and it provides a sense of proportion, compensating for all the risks you take. My job is about place and the things that are connected with it. You see the landscape through many people's eyes."

Born in Birmingham, where her grandfather worked in the jewellery quarter, Helen lived in Halesowen and Leamington Spa during her childhood and has since lived in many different places around the country. "The constant feature has been always coming back to Swaledale for holidays," she says. "And then when we moved here it was the first time that I have felt a real sense of home and belonging. I really love coming in on the bus when I've been away down in London or just out for the day and, as we approach Reeth and I see it nestling in the distance, I feel like I am coming home. It's a lovely place to live. There's an idea about Reeth that it is an isolated place but I can be at King's Cross in three to four hours." And living in a small village doesn't mean there is a limited social life. "I'm busier than I have ever been," says Helen. "I could be out every night if I wanted to, there is always something going on - films, book clubs, history society, bridge. And at the museum we are providing quite a lot of the social life of the village. People have hired the building for birthday and anniversary parties - it's a place to be."

Helen and Alan don't drive so they use public transport and the local community bus to get around. "It's a great service and it works well for us but it does mean that you have to plan ahead," says Helen. "You can't just suddenly decide to go somewhere, but the advantage of a small community is that if anyone is going somewhere they will pop in and ask if you need anything. People are amazingly helpful and generous, so it normally works out."

The generosity of people in the community and further afield extends to making donations to the museum fund. It is easy to understand how visitors to the museum might be moved to do that if they have met Helen and witnessed first-hand her obvious love of, and commitment to, the building and its artefacts. "We have had money donated from all sorts of sources," she says. "There is a couple from Brixton who come to stay with us every year with a friend of theirs. Sadly he died in a road accident and we were very touched to hear that he had left us two thousand pounds in his will. A lady who had family roots in the area died recently and her family generously asked that, instead of flowers, people give donations to the museum. It's that kind of thing that makes you want to go on. If there are people putting their faith in us, you have to." It is reassuring to know that this very special place in a beautiful corner of Swaledale will continue to be nurtured and cherished for others to enjoy for many years to come.

www.swaledalemuseum.org

9

*

Isobel Davies

Ethical entrepreneur, Swaledale

"When I started out, my main aim was to try and revitalise the British textile industry," says Isobel Davies, founder of Richmond-based ethical fashion company Izzy Lane. "It used to be such an important industry in this country - particularly in Yorkshire." Isobel launched the company in 2007 and it has since become one of the leading ethical brands not only in the British fashion industry but throughout the world.

Using the wool of the six hundred rare breed Wensleydale and Shetland sheep that Isobel rescued from slaughter, Izzy Lane creates luxury woollen garments which feature regularly on the fashion pages of glossy magazines such as *Marie Claire, Cosmopolitan* and *Harper's Bazaar*. The range of clothing includes skirts, dresses, scarves, cardigans, coats and jackets and the collection is one of the highlights at London Fashion Week every year. The company has won awards both for design and for animal welfare including the

Isobel modelling some of the Izzy Lane range

prestigious RSPCA Good Business Award in 2008 and, in the same year, the RE New Designer of the Year Award at the world's first ethical fashion awards in London. At the root of the company's ethos are three strong driving forces: animal welfare, traceability and sustainability. "When I first started researching into setting up an ethical fashion brand there was no traceability in the fashion industry," says Isobel. Through her efforts, however, this has now changed significantly and many fashion students today study the business model she has created. "We have quite a few students getting in touch who have learnt about Izzy Lane on their course," she says. "The good thing is that in a few years' time they will be working in the industry, so it gives you hope for the future. I have been asked to speak to first year fashion students at the Royal College of Art to inspire them. I think it is important to do it because those students are the building blocks of our future and that next generation is going to be crucial in the way we develop."

Isobel was born in Nottinghamshire but moved to Yorkshire with her parents when she was eleven. She left home at the age of seventeen, first working in France and then in London in the music industry, but says that she always considered the North 'home'. "I started off playing the saxophone and then I moved into singing and songwriting, touring around Europe with various bands, but I always used to come back to Stokesley where my parents lived." She eventually gave up performing - "I had a terrible singing voice" - to concentrate on songwriting. She was becoming established and successful in this area when, in 1994, she had the idea of setting up an organic fruit and vegetable box delivery scheme. She canvassed opinion in her local area in South London, the response was overwhelmingly positive - and Farmaround was born.

"I thought my future was going to be as a songwriter,"

she says. "I never intended to stop the music but Farmaround just took off and took over and I never looked back." The scheme, which was the first of its kind in the UK, is still going strong, delivering all over the country with Richmond as its northern base - and Isobel is still very hands-on, handling all the importing and distribution herself. She has recently launched a new website for Farmaround which includes a 'Pick your Own' option giving customers the choice of which vegetables they receive in their box so they can make a bespoke order.

It was through her contact with farmers up and down the country while she was sourcing organic fruit and vegetables for Farmaround that Isobel discovered how much wool was being wasted. "I met a lot of farmers who told me they were either burning the wool or paying someone to take it away," she says. "It was the wool merchants who were making the money and the farmers were subsidising them. It has been a vicious circle with British wool: because they are not getting a financial return on the wool, farmers have not been inclined to keep sheep that have good wool quality. It's a shame really because a lot of our indigenous breeds are good for both meat and wool. Only a few decades ago, the wool cheque would pay for the farm tenancy for Dales farmers." Isobel had been thinking for some time about setting up an ethical fashion company and this coincided with her growing desire to leave London and head back to the North - she eventually moved to Richmond, partly to be nearer to her parents, in 2003.

Unlike most other British woollen goods, where the provenance is not known, Izzy Lane's products can be traced directly back to wool from Isobel's flock. Due to commercial pressures, most sheep do not live out a natural lifespan - they are either slaughtered young for meat or as soon as they are no longer useful for breeding. "I realised

Isobel feeding some of her Shetland sheep

quite quickly that the only way I could get my project off the ground was to have my own sheep," says Isobel. "So I started 'rescuing' sheep. I started off with four lambs and I asked a local farmer if I could keep them in his field." The six hundred Wensleydale and Shetland sheep Isobel now looks after, with the help of the same local farmer, Ernest Ayre, will live out their natural lifespan at the sheep sanctuary in the fields near her Richmond home. Wensleydale sheep live for eight to ten years on average while the hardy Shetlands can live up to eighteen years, so it is a long-term commitment and responsibility. "There is nobody else that I have found keeping sheep just for their wool," says Isobel. "But what we are doing raises the whole notion of traceability. The great thing about the internet is that people have heard about what we are doing and it gives them ideas to do something similar."

The company employs fifty or so hand knitters around the Yorkshire Dales who make garments for on-line sale both in this country and abroad. "We are relatively small producers," explains Isobel. "We make about a thousand garments a year - but we have a very loyal customer base. When I first started, we knitted masses of stock but now we more or less knit to order." In keeping with the sustainability principles of the company, local skills and knowledge are used wherever possible. "We started off using a worsted spinner in Bradford - the last one in the city - who, unfortunately, has since gone out of business. We now use one in Halifax and our weaving is done up in the Scottish borders. We are also working with a new manufacturer in Hawick who has hand-framed looms." The products are popular with customers from all over the world. "We get

orders from the US, Italy, Australia, New Zealand, Dubai. And it makes more sense for us to sell this way rather than retailing in shops because we are a niche market and only have a finite amount of material." The sheep which provide that material are sheared once a year - the Shetlands, which make up eighty percent of Isobel's flock, are sheared in early June and the Wensleydales in July. "The wool of the Wensleydales makes beautiful knitting wool and the Shetlands' we weave into cloth," says Isobel. "We have to shear the Shetlands quite early because otherwise they shed their wool automatically. You have to wait until the fleeces rise from the skin - Ernest seems to be able to tell when they are ready. The Wensleydales are interesting because their fleece is more like hair - they can actually get sunburnt."

In the autumn of 2011, Izzy Lane had the opportunity to gain recognition on the High Street when the company was approached by Top Shop to design a capsule range for sale in their stores. "We created three coat designs for them last autumn and this year we are doing a special range of luggage and knitwear for them," says Isobel. "To get onto the High Street has been brilliant because it allows us to reach a wider audience and gives us a platform to go global." Indeed, as a result the company was invited to the Milan Fashion Show for the first time. Isobel is involved in the design of all the garments but is also very open to fresh ideas, working with young designers. The latest collection is, she feels, the definitive Izzy Lane. "I'm really pleased with it. I think it's taken quite a long time to pinpoint what the brand should look like design-wise - it's been a gradual evolution, really - but this collection absolutely captures it. I look at the clothes and think I would like to wear all of those! It's been a sort of trial and error. Going forward, we are aiming to make a more focussed and smaller collection: catering for a niche market, it makes sense to keep things small. We are going to focus on coats, jackets and knitwear fairly evenly split."

With the success of Izzy Lane and the fact that others have taken up her call to action to put ethical fashion firmly on the agenda, Isobel feels as though she has achieved a measure of what she set out to do when she first set up the company. "It's really impossible to know whether you are ahead of a trend or whether you instigated it but, in a way, I feel like I have done my job," she says. "I don't feel like I have to be concerned about the British textile industry any more. I hope that what I have done will give people confidence to invest in manufacturing, train people back up again and expand the industry to meet demand."

Isobel's most recent venture is Good Food Nation, launched in November 2012 in partnership with *Daily Mail* columnist and fellow animal welfare campaigner Liz Jones. Comprising Cow Nation and Hen Nation, they are providing a range of cruelty-free milk and eggs on sale in Selfridges on Oxford Street in London and online through Farmaround. They are working with two farmers, one in North Devon whose dairy cows are providing the milk and another farmer in Sussex whose hens are providing the eggs. Most chickens are slaughtered after seventy-two weeks when they have reached their egg-laying peak, while dairy cows are slaughtered after about five years. "All the cows and hens will now live out their natural lives," says Isobel. "It's still on a fairly small scale until we get some proper distribution through supermarkets, but Marks and Spencer seem quite keen and Selfridges will continue to sell our products. We are starting to develop cheeses and yoghurts; and we are thinking of making cakes too because as hens get older their shells get very fragile so their eggs will have to go into processing. Selfridges have said that they would take any product that we do, which is great. With Marks and

Spencer, if they went for it, they would want products in two hundred stores minimum, which would mean we would have to increase production. That would mean proactively trying to find other dairy farmers to produce for us, but it is a viable business model - there aren't many small dairy farms left and we are offering a lifeline to those who want to go down that route. This is not a product that we can take lightly with the keeping of the animals for life - we have to know that there will be sustained demand for our products and that it isn't just a fad."

The feedback from the launch was very positive, with Isobel receiving emails from people all over the country in support of the idea. "We have had a fantastic response. I think something like this emotionally bolsters people who do love animals - it makes them feel that there are other people out there who care. It brings people together." Isobel's commitment to animal welfare has been a life-long passion and she continues to campaign and raise awareness of the issues that matter to her through her businesses while remaining pragmatic, non-confrontational and non-judgmental.

"I feel a responsibility to let people know about the issues and to keep talking about it. Most people just don't know. Good Food Nation gives us a platform to talk about it - most people don't know that egg-laying hens are slaughtered after seventy-two weeks. Even if they don't support what we are doing, it's good for people to know. I had an expectation when I was younger that people would eventually stop eating meat - I think I was quite naive. I thought that we would evolve to become more sensitive but actually I think that in my lifetime it has got worse. The pressure that supermarkets have brought to bear on farmers and the knock-on effect that has had on animal welfare has been quite damaging. In the old days a traditional dairy farmer would have around a hundred cows and he would know each individual animal. Now there are some farms of around eight hundred cows and you just don't have that sort of relationship any more - but what are farmers supposed to do when the pressure is on them?"

Isobel manages to divide her time between her three businesses sensibly but her involvement in each is flexible depending partly on which needs most of her attention at any given time. "In a way it's a constant battle because you only have so much time and energy," she says. "What I find is that my emphasis and my energies tend to shift. It's really hard to be full-on with all the businesses all the time." She spends most of her time in Richmond, only going down to London if there is a specific reason to go, and she is very happy living in the countryside, although she thinks of

The falls at Richmond

herself as someone with a mixture of town and country sensibilities. "I am sort of a hybrid," she says. "I like the mindset of city people but at heart I am a country person. I think I need to be in nature. I am out in the countryside for about an hour or two every day - walking my dog or driving out to different parts. There is a walk I really like at Marske down by the river. There is so much history in Richmond and it is so unspoilt. I get a lot out of being in this environment when I am here and I would find it hard to live anywhere else." She is aware, however, that being away from London can mean that she is out of the loop in terms of being up to date with the latest developments. "In business you have to keep your finger on the pulse and I have been quite good at being ahead of trends, so I do worry a bit that being up here I am not getting that information to be able to make those judgments. Having said that, these days it's a different world with social media and the advances in technology. Fifteen years ago I could have been quite lonely living up here on my own but now you can be connected around the world."

Isobel feels that her experience in the music industry helped her to approach business in a different, more creative way and that has continued to feed into her current ventures though she admits that she does sometimes think back to her songwriting days with nostalgia. "I have a strange relationship with music - I can't just listen to it as a background. I get drawn into it and I take it too seriously. I miss it when I do listen to it - it makes me really want to sing. But I have found another way to express myself in the meantime. I do feel like I have got unfinished business with music, though, and one day I would like to start writing songs again - but if I did write a song there would need to be a point to it. When I was in the music industry I felt quite marginalised, I didn't feel part of mainstream society so when I started up Farmaround I remember how good it felt. When you are being super-creative, it's all so self-centred. It was so nice that Farmaround wasn't about me, it was about something bigger - the environment and greater causes."

Having already achieved so much by the relatively young age of fifty, Isobel still has one unfulfilled ambition. "I always thought I would be a writer when I was younger," she says. "So I would like to write a novel - based on myself inevitably - and it would be great to have a go at writing a sitcom. I have always felt that life is short - I don't have any children so I feel that any impact I have is what I do in my short life. I am in a position to do these things and I am a risk taker; I'm always prepared to go further. At school I was like that - I was always the first to be letting off the fire extinguishers or running off during cross country. When I went down to London I met other people who were prepared to go further. I have always wanted to do things that haven't been done before."

That pioneering spirit has meant that Isobel has had the courage to create an admirable legacy that will continue to have a positive impact for years to come. "All my businesses feed in to my vision of the world - they are a vehicle for me for the things I really care about," says Isobel, who insists that there has been no grand plan in her business success. "However, when I look back at my childhood, I can trace it all back to then. I always had a shed full of animals that I had rescued - hedgehogs and newts and so on - and there was also a bit of a 'wheeler-dealer' side to me. As a teenager I used to breed rabbits for sale and sell bags of manure from the riding school I went to." She is driven by her passionate beliefs rather than ambition. "I had wonderful parents and a very happy childhood and I've been lucky enough to have had a harmonious life with people so in that sense I don't want to be radical or over-provocative or judge what other

Isobel modelling one of the coats in the Izzy Lane range

people do - but I am doing this to make a difference. I just wanted to create a business model with ultra high animal welfare principles that others could replicate. I've never been interested in business for the sake of it - I can only get enthusiastic about a business that I feel I really care about."

www.izzylane.co.uk

www.farmaround.co.uk

www.goodfoodnation.co.uk

10

*

Kimberley Brereton

Community publican, Coverdale

Kimberley Brereton hasn't lived and worked in the Dales for very long but, in the short time that she has, she has made a considerable impact, particularly in the village of Carlton-in-Coverdale where she has recently taken over as the publican of the newly-reopened community pub, The Foresters Arms. She and her husband Allan were regular visitors to the Dales and stayed as guests at The Foresters while on a walking weekend in the Penhill area, so they were well acquainted with the premises when the possibility of becoming its landlords arose.

In 2011 they came to the Leyburn Food Festival and, as luck would have it, happened to see the stall advertising the details of the Coverdale Community Pub project. They chatted to Peter Dinsdale, who was on the committee, and, since they were involved in the pub trade at the time, offered to help out in any way they could, if required, and made a pledge to the project. They later received an email about how it was progressing which explained that the committee were now looking for potential tenants. "We had been thinking about running a business up here - we had already looked at a coffee shop in Richmond and a B&B in the Dales, so we decided to apply," says Kimberley. "I was an area manager for a pub company, Barracuda, looking after twelve pubs from Nottingham to Middlesbrough and out to Grimsby. Allan and I had also run pubs in our twenties and thirties for about eight years when the children were small but that meant moving around a lot and we wanted to settle in time for them to start their secondary education. So, we

Kimberley and her husband Allan created their own sample menus

moved to Doncaster, where my parents lived, and I worked full-time as a trainer for Toby Carvery pubs, training all the pub staff." The work, however, involved a great deal of travel - five hundred miles a week was not unusual - and she was becoming increasingly aware of the pressures and controls upon those who were not independent landlords. With their youngest child Lily just coming up to finishing A levels, Kimberley and Allan felt it was a good time to announce their depth of interest in the community project. It seemed ideal to them: a tenancy which was affordable, but with the opportunity to run it as their own business, a free house with no ties to a particular brewery or controlling organisation.

There were eleven applicants for the tenancy in the four months open for making a bid. Each was sent a template on which to set out their business plan. Kimberley and Allan took the process extremely seriously, cancelling their planned holiday in Scotland and visiting a lot of Dales pubs to generate ideas for their own submission.

"We sampled a lot of menus in lots of different pubs," says Kimberley. "We put on about half a stone in weight!" Interviewing procedures reduced the list to four, followed by presentations to the committee which cut down the list still further. "We had to present our vision for the pub and how we would see it running. Our message was that you need the visitors and the people staying in the area to make the business viable but it is also critical to have the support of the local community - the people who come in day in, day out." Kimberley and Allan also cooked, creating their own sample menus so that committee members could taste for themselves the type of food (and ales) they wished to offer in the future. Of course, there are obligations to the community supporters but you can't have a committee of fourteen to run a pub so, in effect, Kimberley and Allan -

who is a paramedic based in Richmond and helps out at the pub when he is not working - have autonomy over how the business is run, what initiatives to try out and simply pay for the lease. "The committee don't have any involvement in the running of the pub - that is our job - and it works really well," says Kimberley.

From the moment she and Allan were chosen as the new tenants, Kimberley has been extremely busy. Right from the start - they moved in on 16th December 2011 and the pub opened on Christmas Eve - she was accommodating shooting parties even though the pub kitchen only opened in March. Prior to that, she was using her own upstairs kitchen provided as part of the tenants' accommodation, coping with seventeen diners on Valentine's Night. Despite the frustration of not having the full facilities available initially, the interim period allowed Kimberley to plan what she would be offering on the menu. "Actually, not having the pub kitchen then was a godsend," she says. "It gave us a chance to get to know the local people and for them to get to know us."

A sad memory for Kimberley of this time is that her father, back in Doncaster, died the day after she and Allan moved to the Dales. He had been in the RAF when Kimberley was a child: she was born in Germany and much of her childhood was spent moving home both in the UK and abroad. She went to boarding school in Reigate between the ages of ten and sixteen and then moved back to live with her parents who were by then based in Lincolnshire. This was where she met Allan, who was in the Navy at the time, and they got married when she was eighteen. Before moving to Coverdale, Kimberley and Allan had been living in the same village as her parents near Doncaster and had rented a house for their elder daughter, Amie, and Lily so that their lives wouldn't be disrupted by the move to Coverdale. Their son, Aaron, who studied at Bristol University, is currently working as a radiographer in a hospital in Exeter. Exchanging contracts on the sale of their house was delayed by a month. The new kitchen was available for the first time on Mothers' Day 2012, which was a particularly busy day for bookings. And since that day they haven't looked back, genuinely surprised by the amount of bookings they have had and how far some Dales people have been prepared to travel for a meal out - from Reeth, Hawes and Richmond, for example.

One of Kimberley's early initiatives - before the pub kitchen was open for business - was to organise a series of monthly traditional afternoon tea events which she advertised in the *Darlington and Stockton Times*. "I collect old china and we put out nice old-fashioned tablecloths and made the food in my kitchen in the flat upstairs. The people who come are mainly older ladies - who would not necessarily usually go to a pub - and we have had up to around twenty-five or so regularly attending." As a further part of her campaign to publicise The Foresters Arms, Kimberley has joined Welcome to Yorkshire. As a consequence she has been involved in networking events which have created helpful links to increase business and useful associations have been made with Wensleydale Galleries and Royals, both based in Leyburn. She was also invited to the preview event for the latest ITV series of *The Dales*, hosted at the Plaza Cinema in Skipton.

Initially Kimberley was so busy dealing with the large number of customers that, ironically, marketing had to take a back seat. Now, Kimberley feels, they need a second full-time cook to work alongside their cook Fiona who already works five days, with Kimberley helping out on a couple of days a week. She employs twelve full-time and part-time staff, some of whom are part of long-established Coverdale

The bar area of The Foresters Arms

families, and some who are more recent arrivals. "I wanted to recruit local people," says Kimberley. "We have a whole range of youngsters who live nearby serving in the dining room, preparing vegetables in the kitchen or pot washing. There is a lady, Lindsay, who had worked at the pub before and another, Alison, who moved up here with her children from London a few years ago." Many are still students earning a little extra money, but others rely on the work. The success of The Foresters Arms is not simply that it is providing such a valuable service to the immediate community and further afield but also offering employment in a location where jobs are scarce.

The quiz nights are proving popular. "The first one was amazing," says Kimberley. "We had about fifty people turn

One of the pub's dining areas

up! We run them every two weeks on a Monday evening and we always get good numbers. We also have an open mic evening every two weeks, which we are trying to build into a bigger event." Earlier in the year, when a local band, The Blethermen, appeared over the Easter period, the pub was packed, coinciding with the very successful Cask Ale festival organised by Allan, with ten different real ales available. Similar events are now planned as a regular feature on the calendar. Interestingly, the Wensleydale Brewery (now based at Manor Farm, Bellerby), whose ales are available at the pub, started as a microbrewery at The Foresters Arms in 2003. Informal music evenings have taken place in the bar area where a local band, White Spirit, have occasionally played. "That kind of event is great," says

Kimberley. "It brings in people who are new to the pub so I'd definitely like to arrange more live music evenings." Kimberley has also organised several fund-raising events for her chosen charity, the Air Ambulance service.

During the Queen's Diamond Jubilee celebrations, the strip of land outside the pub was used for a village street party with gazebo awnings for children's activities such as face painting and 'make a crown' and a scavenger hunt. A local farmer collected chairs and tables from the village hall and dropped them off at the pub along with extra bales for seating. "Everyone in the village was asked to bring a plate of food," says Kimberley. "It was lovely because people were so generous that we had tons and tons of cakes, sandwiches and scones. There was so much food!" On the Sunday of the Jubilee, the Duck Race took place on Mel Beck, with one hundred plastic ducks being released at a time, and with over nine hundred tickets sold. The Foresters is very much at the heart of these communal events, entirely justifying the local commitment to ensuring the pub was re-opened. Every other summer is the Foresters Day March, a grand celebration mounted by the Ancient Order of Foresters - a friendly society - from which the name of the pub is derived.

Kimberley and Allan feel as if they have been welcomed into the village with open arms. The community pub committee members have been particularly helpful and have given their time freely to be supportive. With the wet weather that was such a feature of the summer of 2012, local people made up for the shortfall in the number of visitors who might have been expected to visit the dale in general and the pub in particular. A third of the village consists of holiday homes and second homes: thus, people using such accommodation in a year of normal weather would naturally visit the pub in good numbers.

The name of the pub is derived from the Ancient Order of Foresters

The idea of a 'community pub' might, potentially, be off-putting to visitors to the area who fear they could be excluded but the message that all are welcome is clearly being communicated because forward bookings are looking increasingly healthy. "Before taking the pub on we looked very carefully at where we wanted it to sit in terms of food, ales and wines," says Kimberley. "We wanted it to be a traditional pub with a good quality menu. It's the same with the wine - we have a full wine menu. I want it to be the kind of place where you could come in for your tea on a Thursday evening and also come for a special birthday meal on a Saturday night." Visitors who do seek out The Foresters are always interested in the way the pub is owned and managed - and an increasing number of groups in rural and urban areas all over the country are investigating this way of ensuring the survival of their pubs in the face of hundreds of closures each year.

Kimberley's vision is to surprise people with the quality of what is on offer: the ambience of the pub itself, the nature of the food and the comfort of the refurbished guest accommodation, with three double rooms available. The Foresters combines being a warm and welcoming traditional pub with appealing dining areas and excellent bed and breakfast facilities. "We're not aiming to be a gastro pub but we try to use as many local suppliers as possible for the food and local breweries for the ales," says Kimberley. "Already we've been shortlisted for Yorkshire's favourite pub." And this is all within six months of taking over the premises. Stocking a wide range of bottled and cask ales means the pub has received a commendation from CAMRA. Their meat is supplied by Hammonds of Bainbridge, where they also get most of their eggs - Harold, in his eighties, still does the delivering - while Carricks, near Bedale, supply all Kimberley's vegetables.

There's no disguising that the move to Coverdale from South Yorkshire has been a big change in Kimberley's life, particularly leaving her daughters behind and selling a home. "You are moving home and business in one go," says Kimberley. "And you live where you work so your home life is very open to everybody else; you have to get used to that. In a small community like this everybody knows everybody - and we came from a place where we didn't know a lot of the neighbours. It's a positive change but it takes a while to make the adjustment." Some people have said that it can take forty years to become accepted as a local but others have offered the comfort that by running the village pub you are instantly a local. "If you make a mistake, it's known about. Yorkshire folk are very honest and direct and they let you know what they think - but that's a good thing."

Running a pub, especially a successful one, is hard work. It's full-on and full time. "It's a twenty-four hour a day job. I never get to the end of my list and there's always something to do," says Kimberley, "but I love it. You have to have a rolling list - be flexible. Time passes so quickly because of being so busy. My friends at work joked that I would be crocheting behind the bar, but there's no time for that! I've never been busier." She and Allan appreciate that, eventually, they will need to make time for themselves to enjoy where they're living and to get out exploring the Dales once more. Kimberley has only been on one walk since taking over the pub, and that was an organised event starting from The Foresters itself. "We live in this beautiful countryside and I don't actually see that much of it," says Kimberley. "You have to make yourself do it - step back and enjoy the Dales again." The local solicitor, John, who helped out with licences, advised Kimberley to "just take a deep breath. You are here for a long time - you don't have to do everything straight away." Making friends is also on

A view towards Carlton-in-Coverdale

Kimberley's list of things to do: as she points out, "as a landlord and landlady there are a lot of people who know you, but you need to make time to develop real friendships. I went to a WI meeting with Lindsay not long after we moved here and I would like to go regularly but I haven't been able to so far."

Kimberley admits to being "desperate to get it right" with the pub so she has been very pleased that the feedback received so far has been very positive. However, change and improvements are her constant watchwords so she plans to alter the menus seasonally, advertise to attract more niche tourists, such as cyclists and walkers, and provide a bike rack as a helpful extra. She also intends to improve the look of the outside of the building, adding to the floral displays and creating a vegetable patch where they can 'grow their own' for the kitchen. She anticipates initiating more activities to augment the regular events they are already hosting, including a cider festival, a charity bed race, *Call My Bluff* wine quizzes and themed food evenings. There will be further training for the staff, reconfiguring the kitchen to make it more efficient and there is no intention of letting the activities fade out - Kimberley is well aware that different kinds of events attract different kinds of people, all of whom add to the sense that The Foresters Arms is a thriving success story. "This has to be a viable business," says Kimberley. "Lots of pub businesses are failing."

Those who have invested in the community enterprise receive their dividend from the pub rental. At present, that rental is deliberately fixed at a rate that Kimberley acknowledges is "very reasonable" in order to give them a helping hand in establishing the business, but that rental could increase in due course. They purchased a five year lease with an option to renew for a further five years: this lease can be sold on but any 'goodwill', that is to say any building up of the customer base, belongs to the community shareholders. Both Allan and Kimberley are very happy with the arrangement since the idea of living in the Dales has always appealed. A good local pub is the heart of any community offering a warm welcome and a cheering atmosphere - a home from home. Clearly, Kimberley and Allan feel very much at home at The Foresters, a pub where all are made welcome, and it is not just the villagers of Carlton-in-Coverdale who are extremely grateful for their arrival.

www.forestersarms-carlton.co.uk

11

*

Moira Metcalfe

Artist, Wensleydale

Moira Metcalfe is one of the most successful artists currently living and working in the Yorkshire Dales, though it has taken time, determination and considerable natural talent to achieve such recognition.

"I think people in the Dales, women in particular, have to make their own employment in some kind of way because there isn't much paid work available here apart from farming and tourism," says Moira, who lives in the village of Appersett. In fact, she has developed a sizeable niche for herself over the years, driven by her creative impulse and artistic vision, which have combined with her love of the local landscape to positive effect - her distinctive style and bold use of colour is greatly and widely admired. Moira works mostly in oils and her images of field barns - one of the most recognisable features of the Yorkshire Dales landscape - and fells blur the boundaries between figurative and abstract art in an exciting and original way.

Moira working on one of her abstract paintings

In her sizeable lounge/studio, which also now serves as a weekend tea room for visitors to her gallery, she is surrounded by her oil paintings and prints which hang from the walls, or stand on easels or lean in stacks against a nearby support. Her work ranges from distinctive and colourful landscapes to bold, sweeping abstracts inspired by the contours and special light of her very particular part of Yorkshire, a little to the west of Hawes, in the green delight that is Upper Wensleydale. Light is all important, both when she is out and about in the landscape itself but also when she angles her easel close to the north-facing window of her working space - she finds the light from the north to be more clear, consistent and untainted by shadows than from any other direction.

Moira's work has recently acquired a less figurative

edge, which she acknowledges is almost certainly connected to the changes in her domestic circumstances: after a long marriage and with three grown-up children, she now describes herself as 'nearly divorced'. "My work has become much more abstract and personal," she says. "Some of it is quite dark, but it has helped me get through what has been a difficult couple of years - and the work has been very well received so far."

Moira's parents moved from Leeds to Appersett at about the same time as she completed her college studies in Fine Art - she followed a Foundation course in York and then a three year degree at North Staffordshire Polytechnic (now Staffordshire University). "My father was working between London, Belfast and Leeds so he did a lot of travelling. He was the sort of person who liked to travel the back roads and he came through here and thought it was a lovely area. He saw this house was for sale and he decided to buy it." Her father continued to work all over the country but the decision to live in a relatively isolated part of North Yorkshire was typical of his refusal to conform. Moira's father worked in marketing and management but was also very musical (he was "an extraordinarily good pianist" says Moira) - perhaps an influence upon her own artistic inclinations.

She met her husband-to-be, Bryan, a local sheep and cattle farmer, in 1979. "When we got married we needed to find somewhere to live and we moved in to one side of the house - my parents lived in the other side. I have been here ever since," says Moira. "I raised a family here and I have been painting full time for about twenty-five years. I have a son and two daughters: Spencer is head chef at a local hotel; my middle daughter is married and lives in Chester - she did fashion and textiles and runs her own business from home; and my youngest daughter lives in London and works in marketing."

The house is called Netherbar but the gallery room is now known as Artbar. Bryan's farm, run along with his two brothers, extends for eight miles within Wensleydale and towards Garsdale, raising both sheep and cattle. As a farmer's wife, Moira became involved in helping out with lambing in the spring and with gathering in the hay in the summer but essentially she was, even at this early stage, already forging a career as one of Yorkshire's most talented and sought-after rural-based artists. By the time the youngest of her three children was attending school, Moira was very much back to tackling art projects in whatever free time remained to her, "though at first, I couldn't believe how rusty I was," she says. At one time the sale of her paintings was seen as a valuable contribution to the household budget: today, in her changed circumstances and in a different phase of her life, it is her main income.

Moira's artistic influences relate very much to the environment she finds herself in which is why much of her early work (beyond art school) reflects her awareness of the surrounding landscape, with the nearby hill Stag's Fell an abiding presence both as a view from her garden and in her painting. Interestingly, though she still continues to work mainly with oils, Moira's return to abstracts, which are now a dominant aspect of her current output, is a continuation of an earlier fascination with the non-figurative. "When I was at college, I worked in a range of media," she says. "I did a lot of abstract and three-dimensional work - I even did sculpture and welding - and those are the same elements that fascinate me now. Abstracts are a very personal expression and are actually much more demanding to do than the landscapes. There is an awful lot more care and thought that goes into it. It's almost like you have to serve a kind of apprenticeship in figurative art first before you start to move on to abstract work," she explains. "It feels like progress for most artists and it becomes that bit more distinctive." Comparing her recent output with her earlier work, it is possible to see a new complexity, richer colour and a different form of artistic expression. "Some of my abstracts look quite dark but actually they have a lot of light in them," she says. "I use the colours to depict the kind of mood I want to convey."

Moira also earns a living as a tutor in art in courses she runs at the Stone House Hotel, near Hawes, and she has recently branched out into freelance tutoring, offering guests residential two-day courses based at Artbar throughout the year. "It will be an opportunity for them to switch off from everything else and immerse themselves - just think and talk about art as well as do a lot of painting," she explains. Moira can offer breakfast, lunch and home-cooked evening meal (or a visit to one of the local pubs) and tuition - as well as some free time - at a competitive price (roughly three

Bridge at Appersett

hundred pounds) with up to four students at a time. Interaction between fellow guests is important, she feels - and one-to-one tuition can be quite exhausting for both parties. Moira believes there is a lot that beginners can discover in themselves as artists with awareness of technique, composition and perspective, especially when working with a forgiving medium such as pastels or charcoal. "I'm surprised at the number of people who try to start off with watercolours," she says. "It is actually one of the most difficult areas of art to master. If someone has enough raw talent it will come out, but the main thing is for people to produce something they are happy with. I find it very satisfying to encourage that kind of creativity."

Through the Swaledale Festival, Moira became involved in a fascinating venture in which she collaborated with and mentored a schoolchild and both she and the child produced a piece of abstract art work which was jointly sold at auction in Richmond. Moira was extremely enthusiastic about the project. Her desire to encourage and nurture new talent is very apparent.

Much as she enjoys tutoring, it is, naturally, Moira's own work to which she gives greatest priority. She now finds her work being exhibited more and more widely - there has been a recent exhibition in Marylebone, London, as well as in Geneva in Switzerland. In April and May of 2012, Moira exhibited her work jointly with highly respected landscape photographer Joe Cornish. Cleverly titled *On Common Ground,* the exhibition compared and contrasted the techniques of landscapes on canvas and in photography. "Exhibitions can be quite demanding, but it is great to be asked - especially when people seek you out," says Moira (as was the case with the Swiss exhibition). The opportunity to show her paintings at locations such as the Wensleydale Creamery in Hawes guarantees that in the region of a quarter of a million visitors each year will see her work. The newly-refurbished White Hart Inn in Hawes exhibits a specially commissioned landscape oil painting by Moira featuring the distinctive peak of Stag's Fell (visible to most Hawes residents when glancing northwards) and a small white hart whose final position in the finished work took several attempts to determine. She is often commissioned to paint particular views and she says she now also has some regular collectors of her work.

Moira's technique relating to her own figurative work is to take her camera and sketchbook into the surrounding countryside and, where necessary, conduct some research into the particular feature of landscape that has caught her attention. A great deal is kept in her head rather than captured on film or on paper - images and ideas that feed into both her finished landscape and abstract works. She has taken her easel out into the countryside on occasions, "but it's a nuisance when the midges get stuck in the oils." Also, there is a very real likelihood of crowds gathering when Moira paints outdoors, though she is not distracted by the attention. "Often I just go out into the landscape to spark ideas and I will see something that appeals - unless I am doing a commission, of course." However, the vast majority of her painting takes place back in the studio where she assimilates a landscape, allowing her creative interpretation plenty of scope.

Unsurprisingly, Moira's favourite view is from Artbar's doorstep, looking towards Stag's Fell, which she enjoys seeing in different lights and seasons. "It is ever-changing," she says, almost awe-struck by its magic. Just as the natural light in her studio is all-important, her appreciation of the effects of light upon the landscape is particularly inspiring, especially when the effect can be quite striking, as when a beam of sunshine penetrates storm clouds. "I love dramatic

Moira outside her studio

light. It's a split second sometimes and then it's gone again. I have to hold in my head what I can see in that split second. I like strong contrasts - I don't really like painting on sunnier days. When I am doing figurative landscapes I try to adhere reasonably close to what I see so that people know where it is - although the abstracts and the landscapes are getting closer and closer together." Though she never attempts to reproduce a quasi-photographic impression of a view, the shapes and contours of a particular location will often be instantly recognisable while Moira's inventiveness with mood, skies and colours add strongly abstract qualities to her vision of a given scene.

"A painting might take me two or three days to complete - or it might take two to three weeks," says Moira. She enjoys spontaneity but is aware that something special can emerge purely from working and re-working a subject. In ideal circumstances, she paints every day but seizes the opportunity of good weather to be out taking photographs of interesting locations. She loves to garden in the morning for an hour or so as a way of clearing her mind and preparing to work. All the time there is natural light flooding into the studio, she tries to take advantage of the opportunity to paint. The consequence is that the winter season reduces her working time but, as she says, living in this part of the world you have to adapt and work with the seasons. "There is a relationship with the landscape when you live in it. Local people are very aware of their surroundings. You have to find a way of working that allows you to be able to live here. Much of that has to be seasonal - especially when we rely so much on tourism in The Dales."

Moira's weekend café opened for the tourist season of 2011 and has been bringing in many more visitors to Artbar. "Some people feel a bit intimidated by galleries," says Moira. "So I thought that by creating a relaxed atmosphere, with refreshments, it would help people to feel more comfortable in the space. It works quite well." Moira keeps the menu very simple, offering tea or coffee and homemade scones, while the customers can browse the pictures and prints at their leisure. As well as a few tables and chairs inside, there are tables in the garden for when the weather is good - and visitors like to see the peacock and white peahen which have become star attractions alongside Moira's free range hens. The increase in sales of Moira's art work as a result of this initiative has been very noticeable.

What the future holds for Moira is more of the same - being very productive and seeking out further opportunities to exhibit widely. She enjoys meeting fellow local artists at exhibitions, visiting studios and both offering and receiving feedback about each other's work. It wasn't always so.

Moira in the gardens of her Artbar gallery

"When I was at school doing A level Art, I did Architecture rather than Art History because I had a horror of being influenced by established artists," says Moira. "I wanted to find my own way. All my Art teachers thought I was crazy. However, I have always tried not to be influenced too directly in my work." As her own art has matured and her particular style developed and become truly distinctive, she is now much more relaxed about viewing other artists' creative output. "Artists I admire are Turner - because of his use of light - and I love David Hockney's work and his use of colour. I do love looking at other people's art now."

Partly because of having gained her financial independence through her artistic success, Moira says she feels no need to seek out a new relationship. "Women are getting more independent and stronger; it's interesting

A view of Stag's Fell from Moira's garden

having found myself becoming a divorcee at this stage in my life. I know a lot of women who are in a similar situation and I don't think I need to remarry; being totally independent is a lot more important to me." She has an active social life with a group of women friends who value the opportunity simply to be themselves. She likes to eat out and enjoys cinema, theatre and musical events - so Richmond has become a popular venue for her with its Georgian Theatre and The Station arts centre offering a variety of cultural entertainments.

Moira lives in a very small community where the villagers generally get on well with each other, though she also socialises with Hawes folk and with friends in other small villages nearby. "You do need the independence of driving a car if you live here," she acknowledges. However, the Little White Bus (partially-funded and run by the Upper Wensleydale Community Partnership with volunteer drivers) has made a big difference for local people who don't necessarily have their own transport - it provides a link with Garsdale station which, in turn, links Leeds and Carlisle. Moira's mother, who still lives next door to Moira, is well into her eighties and has only just stopped driving. She is now much more dependent on Moira, "but that would be no different if we were living in a city, really," she adds.

"I love living here. I couldn't live somewhere like London. I'm much happier here living with the so-called disadvantages. The advantages are obvious: fresh air, peace and quiet, an open fire, cats, hens, the whole way of life. People say hello here," says Moira. Modern developments have also made living in a Wensleydale hamlet more amenable. "The internet has made a big difference and I do most of my shopping online." For Moira that means not just her weekly groceries but all her art supplies. "Life in this part of the Dales can be tough, and winters can feel long and

lonely, especially for the elderly," says Moira, "but you do adapt to it and learn to cope - or more than cope. You learn to make use of each other in a sociable, friendly way: if someone is going out shopping somewhere, it makes sense to offer to get something for someone else too." It simply amounts to looking after each other, looking out for one another.

After thirty-four years of living in the same location, Moira is rooted in her landscape. As well as having captured it colourfully in oils, it has been transmuted imaginatively into her abstract expression. Her work is as organically connected to her environment as a farmer's. She belongs. "I have friends here. I have a life here. I have made a life here."

www.moirametcalfe.co.uk

12

*

Pat Thynne

Organisational change consultant, Grisedale

"I was desperate to live somewhere like this, so I had to find some way of being and working which would allow me to do it," says Pat Thynne who has been living in Grisedale - one of the wilder and more remote areas of the Yorkshire Dales National Park - since 2004. Grisedale famously featured in a Yorkshire Television documentary in 1975 made by producer Barry Cockcroft who two years earlier had 'discovered' Hannah Hauxwell.

Entitled *The Dale that Died*, the film focussed on former miner Joe Gibson, originally from the North East, who had moved to Grisedale in order to begin a new life as a tenant sheep farmer. The dale had once been the home to fourteen families but by the mid-1970s Gibson - along with his wife and son - was the only farmer living and working in the dale. Theirs was a hard life and the film repeatedly referred to Grisedale as having 'died' in the sense that the small population had moved away from the valley and the deserted houses were gradually becoming derelict. Today there are just five permanent residences in Grisedale including Joe Gibson's farm, now lived in by his grandson Matthew.

Pat's house, Reachey, was a ruin when she first saw it in 2001. "Even so, I knew immediately that I was going to live in it one day," she says. "I think the house chose me; I felt like I belonged here. It took me over a year to buy it and then two years to get planning consent." She rented another property in the dale while the work on Reachey was carried out and finally moved in to her new home in 2006. A qualified lawyer, Pat knew that in order to stay in Grisedale she would need to make changes in her working life. "I couldn't continue being a solicitor and live here, so I had to find a different way of being. That, combined with happenstance, meant that I slid into becoming a consultant."

Pat outside her renovated house in Grisedale

Pat's area of expertise is organisational change, change management, training and coaching, dealing mainly with local government. "What I like doing most is working with management teams to get them to approach things in a different way and then watching the energy develop within an organisation. The more managers listen to and trust their staff, the better. If you give people what they most want in the world - which is autonomy - and trust them to do the right thing, in my experience they don't decide to do the wrong thing. Most staff are not unrealistic; they know what is manageable and achievable. I am a profound believer that happy people work harder. The mental image I have of my role in these situations is as the person in front of a curling team who is rushing about polishing the ice - I am an enabler."

Using Grisedale as her base and travelling to work on

A view of Grisedale

projects locally and all over the UK, Pat has been able to build her consultancy over the past few years and says that living off the beaten track has not been a disadvantage. "Very occasionally the weather gets in the way or, from time to time, the inadequacy of the train services, but I have only missed two appointments in the last eleven years because of the weather - I suppose it's more the fear of it than the reality. And, as you often hear elderly residents of the Dales say, the winters aren't what they used to be. I have met several people who were born and grew up in Grisedale and come back to visit regularly - and they tell tales of winters when the snow was so high that they had to walk to school on the tops of the dry stone walls."

Pat's form of transport is a four-wheel-drive vehicle. She says that since she lives on her own she needs something that will definitely get her home at night. The track down to

Reachey from the only road through Grisedale is steep and can be difficult to negotiate in an ordinary saloon car - although the postman manages to deliver every day. "You have to have a car to live here," she says. "There is no bus service, so I can't opt to take the bus. I have to think about the price of diesel and I try not to be too carbon polluting, so I make sure that when I go somewhere in the car I have at least five different reasons for going. Sometimes I don't leave the dale for a week because there is no reason to go anywhere." There is no popping out for a pint of milk when the nearest shop is nine miles away, so that means planning ahead. "In October I have three tons of logs delivered for my stove which keeps me going through the winter," she says. "I have a wood pellet burning boiler, so I also have wood pellets delivered in large quantities. If you live in a wild place then you have to adapt your life to the weather - I keep a good store of food and fuel and when it is really snowy I just light the stove and relax. I do have a couple of toboggans if things get really bad in the winter!"

The capacity for self-sufficiency is essential in such a remote area and this is something that Pat has embraced in adapting to life at Reachey. "I like the feeling of being self-reliant and I have had to become more practical since I've been living here," she says. "My boiler, because I wanted to make the house as 'green' as possible when I was renovating it, is a bit of a prototype. It broke down recently but, before getting someone out to look at it, I took it apart and put it back together myself first." Her renovation of Reachey was very hands-on and carefully thought out, working with a local architect and craftspeople, and with constant reference to the Yorkshire Dales National Park's Conservation Officer. "One of the arguments I used when I was applying for planning permission to restore the house was that I wanted to honour the old but incorporate the new."

And this is exactly was Pat has achieved - creating a home that looks to the future while at the same time remaining firmly rooted in its past. For Pat the restoration was not only a satisfying architectural project, it was also a very personal one that helped her through what was a difficult time in her life. "My partner, Gill, died in 2000 and we had spent a number of years doing up houses. When I came here in 2001, renovating the house was part of a grieving process. Gill was an artist and she worked mostly in wood, metal and stone - and I wanted the house to reflect that. I also felt a great sense of the house's history and the people who had lived here before - I wanted to honour the authenticity of the building and the fact that it had grown out of the land. I didn't want it to look like an 'alien' in its environment." One part of the building had largely collapsed, so this was the space which Pat chose to reflect the future. "The process of working out what it should look like was fascinating. It was about fitting in with the place itself. I wanted to acknowledge its past and make sure that it wasn't a 'pastiche'. The architect came up with the idea of floor-to-ceiling windows as a kind of echo. It seemed to be a dialogue with the house itself."

While carrying out the renovation, Pat had help from a local councillor in researching the origins of Reachey and discovered that at one time the farmhouse and byre had been home to a family with twelve children plus a farm worker. "I do feel slightly self-conscious about how much space I have compared to the people who used to live here," says Pat. "They had no electricity and got their water from the beck; it would have been an incredibly hard life. The Thompson brothers who used to live in East House as children many years ago come to visit the dale regularly. When they were growing up they kept chickens at Reachey which was a ruin by then. They really like what I have done with the house and that means a lot to me." The restored house is a part of, and is affected by, the landscape in which it stands. "It works with the seasons," says Pat. "In the winter I tend to live in the older part of the house because it is warmer - it's cosy and protective - and then in the spring I move into the newer part of the house which is lighter and I like waking up with the natural light. There aren't any curtains in the house which some visitors find a bit disconcerting. It's about living in harmony with the land. The house is a real mix and complements the romantic and the realist in me."

Deciding how to decorate and furnish the inside of the house has also been a pleasure for Pat and the eclectic mix of North African wall hangings, large oil paintings of Dales landscapes, well-stocked book cases, fascinating objects on the shelves and even a small grand piano in one corner reflect Pat's interests and passions. "This is the first time in my life that I have ever lived on my own and the first time I have been able to decide what I would like in the house and on the walls without having to refer to anyone else," she says. "All the objects have some significance. One of my oldest friends has spent all her working life in Morocco and I have visited her quite often. I love the desert and the colours of the desert. This might sound strange but I feel there is an echo between here and where she lives in the foothills of the Atlas Mountains - the only difference is the sun! The people are living in a marginal place - here there is too much water, there not enough - but people are scratching a living in the landscape." Pat has been trying to cultivate a garden next to the house but the harsh weather conditions of Grisedale - and the resident animals - have made this a difficult task. "It's a real challenge because it's so cold and also almost everything I have planted has been eaten by the local wildlife. I haven't cracked it yet. I would like to be able to grow useful things like trees and food. I started by putting up a deer-proof fence and then, when it was up, I thought, 'That's horrible, I don't want to live in a cage.'"

To live in a place like Grisedale, Pat acknowledges, you have to be happy in your own company but there is also a sense of community among the few residents. "We all get along, possibly because we are not cheek by jowl," she says. "Everybody who comes to live here has a love of solitude. There is a common bond and a recognition that you are here because you like being on your own, but if you are stuck you can go and ask for help. People sometimes ask me if I get lonely, but I think that solitude is different from loneliness. I worked in London for many years and I would feel far more lonely living alone in London."

Pat still has to go to London quite frequently as part of her work but she says that, although she enjoys visiting, she would find it difficult to live there again. The pace of life and the sheer volume of people is not something that she misses. "I was there recently and in the morning when I had left here, as I was coming out of the house, there was a deer running across the field, the curlews were singing and a hare was bounding up the track. Not long afterwards I was in Bloomsbury walking across a park covered with bodies in a very small space. Even though within yards there were several restaurants, cinemas, bookshops and a wonderful cultural mix and diversity, I prefer the silence of Grisedale."

Apart from farmer Matthew Gibson, Pat is the longest-serving resident in the dale which is, she says, unusual for her. "I have lived here longer than anywhere in my life. I have always moved around a lot but I feel settled here." Pat went to school in Worthing and London, and went on to university in Swansea to study Psychology and Social Administration. "When I left school I thought I wanted to be a social worker; after university I had a two-month

A view from Reachey, with ponies

placement in Liverpool in the 1970s. One of the people I was assigned to was an elderly lady living in a prefab and I wanted to try and get her some new furniture. I asked my supervisor and she said, 'You should be teaching her how to budget rather than trying to get her funding,' and I thought, 'This isn't for me.'" Pat went on to work in the voluntary sector in an advisory capacity which is where she first came in contact with solicitors. "I had a pretty working class background and had never really met solicitors before but I thought, 'I could do that.' So, I went to train at the City of London Polytechnic where I met a group of very like-minded people - solicitors who wanted to change the world. Then I trained at Bindmans in London, a civil liberties practice."

Pat's desire to 'change the world', and to combat social injustice, as well as her independence and self-reliance, could perhaps be traced back her childhood. An only child, Pat had an unconventional upbringing, for the time, having been brought up by a single mother. "I never knew my father. I was illegitimate in the 1950s and my mother had to become a housekeeper to look after me. She never had the chance to be the kind of person she could have been. She got a scholarship that she couldn't take up because she had to go to work. She was trapped by class and gender and then, when she had me, by societal prejudice. She worked and brought me up on her own; she never married. I don't know how she did it but I am really grateful that she did. I owe her a debt and part of that debt was to do what she always taught me to do which was to be happy." And it is clear that the life Pat has made for herself in Grisedale certainly makes her happy. "I have the occasional fantasy about doing another renovation because I had so much fun doing this one, or moving up to Argyll in Scotland, which is an area I love, but I don't have a family so my friends are my family.

Some are in Leeds, York and Harrogate, others are in London or on the south coast - and I can get to see them regularly living here. "

Another reason for staying put is that Pat and one of her neighbours, Sally McMullen, have inherited a number of Shetland ponies after the recent death of John Pratt who used to farm in Grisedale. Most of them have been sold but a few are still in the dale. "Sally and I share responsibility for six Shetland ponies and three foals," says Pat. "I had to go on a crash course. I was standing outside the other day grooming one of the ponies and I thought, 'This is not something I ever envisaged I would be doing; this is not a me that I thought existed.' I find that I really like being with them - horses seem to have the power to reflect what's going on within people. They have a great intuitiveness about them. Looking after the ponies is quite therapeutic. Recently I just sat down in the fields and they came and stood around me. One of the ponies put her head on my shoulder and, I know this sounds cheesy but, I had a sense of communion. I knew that she trusted me and that was a lovely feeling. It made me feel good about myself." However, Pat is conscious that she mustn't allow herself to become too attached to the ponies having observed the relationship that country folk have with their livestock. "Watching Matthew and the way he is so matter-of-fact with the animals makes me realise that I don't want to be too sentimental about them - it would be nice to find a balance."

Finding that balance is all part of the learning process that Pat admits she is still going through as someone who has become a countryside dweller relatively late in life. "I am very conscious that I am what is known round here as an 'oft-comer'. I am essentially an urban creature and I have quite well-defined urban skills, so I often ask myself the question, 'Do I have validity here? Do I have a right to be here?' I'm not part of the landscape and I don't have the kind of practical skills that someone like Matthew has. I don't really know how to live in relationship to the environment. Everything is learning and here it's about learning to listen on lots of different levels. I'm trying to get to another level down in order to understand better what this place requires of its humans."

Pat's affection for Grisedale is palpable and it is evident in the care and thought she has put into creating her extraordinary home. She has a peaceful, fulfilled life surrounded by nature, the elements and by people who feel the same way that she does about this small corner of untamed Yorkshire. "The people who live here all have a

Ponies in Grisedale

profound love for the dale. When John Pratt died, his wish was to be buried up at the top of the moor. He has one of the best views of the valley! I personally love Grisedale's silence and its 'almost wildness' - it isn't conventionally pretty like most of the rest of the Dales. There is a real sense of community amongst the few people who live here in that you each have your own space, but you know that you are there for each other if you need it." Perhaps mindful of the fact that Grisedale was once referred to as a dale that had 'died', Pat is concerned about how it - and the Dales as a whole - will survive for future generations.

"The opportunities for work within the Dales are profoundly limited and, with the decline in farming, it's going to diminish further. I passionately believe that, if the Dales are to have a future beyond tourism and retirement homes, it has to find a way to be 'ordinary'. I think a philosophy that honours the old but incorporates the new needs to be applied. We need to think about what happens to the land when it's not farmed anymore. It's not just about how we survive: how else can we treasure and nurture the land? The old way was to impose our will on it but maybe we have to find a new way. It is so precious."

www.reachconsulting.co.uk

13

*

Pip Hall

Lettercarver, Dentdale

"There are around seventy people in the UK earning a living doing lettercarving in stone - and here is a wonderful place to be doing it in," says Pip Hall, sitting in her studio in the village of Cowgill in Dentdale, where she has been living since moving up to the Dales from Reading in 2005.

"In a way I moved up here to be nearer my material," she says. "It was also partly because of a project I was working on but I think I had moved here in my heart already. It felt like a home from home - and it's very beautiful here." Pip's home is what used to be the old schoolhouse in Cowgill - her studio is next to the main house which she is currently renovating. "The school was closed down in the 1960s and it was a holiday home for a while. I will continue its use as a holiday let but I also have plans to run residential art courses here - and turn it into a place of learning again." Pip learned her own craft in her thirties and is rightly proud of being part of the rich heritage of this very specialised profession. "It is an extraordinary tradition," she says. "The skill goes back to one person a hundred years ago, Edward Johnston, the father of modern lettering, who taught Eric Gill: Gill taught David Kindersley, who taught his own son Richard, who taught Alec Peever, who was my teacher."

Born in Romsey near Southampton, Pip moved to Germany as a child, where her parents were teachers in the British Forces Education Services, and then returned to the UK when she was eleven to go to school in Lyme Regis. "After school I did an Art Foundation course in Huddersfield and then, because I had always been interested in lettering, I went on to study for a degree in Typography and Graphic Communication at Reading University between 1983 and 1987. The course at university was great; it was the only one of its kind in the UK and was the vision of one person, Michael Twyman, who created it. We went on study trips to Rome, Florence and Frankfurt. There was a lot of emphasis on letterforms and seeing them in ancient forms in Rome and Florence, but at that time I never thought I would do this."

After graduating, Pip worked in a design studio for six years until the advent of the computer age which changed the nature of design and prompted her to reassess exactly what appealed to her most in the work she was doing. "My training was very much a craft training and I realised that what I was interested in was making things," she says. "I was inspired by a visit to lettercarver Caroline Webb in Wiltshire and it was wonderful seeing her carving really beautiful lettering. I knew then that was what I wanted to do. It was also good seeing it was possible for a woman to do what I had thought of as 'male' kind of work. Stone-shifting, for instance, just not being an issue, with lifting equipment available. So, in my early thirties I decided to

Pip's studio

make a total change of lifestyle. I found a little space where I was living, next to the bathroom, to do my lettercarving – and taught myself. Then I was introduced to Alec Peever who ran a lettering studio in Oxfordshire and worked as his assistant. There I learnt to carve in a much more efficient way and also I learnt how to draw. That is key - it is essential spending time on the drawing itself before you pick up the chisel - and, of course, everything is proof-read at the drawing stage."

With her newly honed skill, Pip went back to work. "A wonderful friend offered me her potting shed as a studio and I worked there for the next seven years." She applied for, and received, a grant from the Crafts Council to set up her own business which included funding for equipment, help with publicity and marketing, and the opportunity to take part in exhibitions. "I showed some of my work with others at Chelsea Crafts Fair and got dozens of commissions: that got things going," says Pip. "I also exhibited at Art in Action at Waterperry near Oxford. There was everything from pottery to dyeing of yarn, to glassblowing - beautiful things being made. These two events were quite crucial in terms of having enough work coming in. I didn't do any advertising and, after a few years, I felt that my business was well established."

Many of Pip's commissions come through the Memorial Arts Charity, based in Suffolk, which she has been associated with for several years. The charity is unique in the UK in that its aim is to support and encourage memorial art and the development of lettercarving skills. To this end, since 1998 it has been running an apprenticeship scheme and lettercarving workshops throughout the UK, raising awareness and appreciation of memorial art as a modern-day art form. They

Pip in her studio

recently set up a national collection of Memorial Arts exhibiting in six garden sites across the UK. One of Pip's pieces, a seat made out of green Cumbrian slate, is exhibited in the garden at Winterbourne House in Edgbaston.

"The charity was set up by Harriet Frazer who started out wanting to find someone to carve a memorial for her daughter," explains Pip. "They are currently concentrating on promoting the collection and developing the training programmes. I got to hear about them through the first booklet they produced and I asked if I could be on their register - at that time I was self-taught. Harriet very politely suggested that I get some training and she put me in touch with Alec. She is a real enabler and supporter. The charity, which celebrates its twenty-fifth anniversary in 2013, has played a huge role in increasing our awareness of lettercarving and the number of people doing it throughout the country, and beyond, and we

The Beck Stone on Ilkley Moor, carved by Pip

are all indebted to Harriet and her team for this continuing support and promotion: it is an extraordinary legacy to be granted to this craft." Pip has contributed to the continuation of the lettercarving skill, passing on the tradition to a new generation, by taking on her own apprentice, Wayne Hart, for a two-year training period funded by the charity. "Because I learned in a studio, I felt it was right to teach someone in my studio," says Pip. "It is very much an apprenticeship way of learning and I feel a responsibility to carry on the practice, to take pleasure in continuing the exploration of lettering as fine craft and an expressive art form."

The process that Pip goes through when she is commissioned to create a memorial piece varies and to some extent depends on the wishes of the clients at what can be an emotionally difficult time. "I had little experience of that aspect of the lettercarver's work when I set up on my own," she says. "Of course, it comes with experience but I hope that I have been able to pass on to Wayne more of an understanding of this very sensitive part of the commissioning process." When Pip receives a commission for a memorial, she usually goes to visit the site where the stone will eventually be placed. "It's useful to see what other headstones are there to make sure it fits in and that does somehow influence the way I design. Sometimes a client will visit me here and sometimes we don't meet at all and it's all done through email - I can take photographs of drawings and they can make changes. Sometimes the first time we meet is when I am going to install the stone. I like using lettering in an expressive way but I do tend to work in a slightly more formal way for memorial stones. It is amazing the variety that is possible even though there are some quite severe regulations and some churches and cemeteries can be quite strict."

The other area where Pip has used her expertise is for a number of large-scale public art projects - her most recent involvement was on the Stanza Stones, commissioned by Ilkley Literature Festival in partnership with the poet Simon Armitage, funded by Pennine Prospects and imove as part of Yorkshire's Cultural Olympiad. The purpose of the project was to develop a long-distance walk from Marsden near Huddersfield, where Simon was born and still lives, to Ilkley, the home of the Literature Festival, punctuated by Simon's poems carved into the local gritstone rock. The route largely follows the South Pennine watershed with the Stanza Stones positioned at six locations along the way - in fact, there is a secret seventh stone awaiting discovery by those walking the entire forty-seven miles of the trail.

"Because of previous projects I have been involved in, I was invited to express an interest. It seemed right up my street - working outdoors and carving directly in the landscape really appealed to me, and poetry has always been part of my interest in lettering," says Pip. "I very much enjoyed working with Simon - I found him incredibly illuminating in the way he talks about poetry. He would make site visits and we would talk about the layout of the lettering together - I always gained some new insight about the poems from our discussions."

By the time Pip came on board, the project had already been in development for a year and most of the locations had been selected by Simon and consultant landscape architect Tom Lonsdale, who also produced the poetry trail guide. "The sheer variety of the locations I found fascinating," says Pip. "At that point they were looking to me to give technical and practical advice and responding to the stones, so I helped to choose the rock," says Pip. "It was interesting that we all seemed to home in on the same places." Each of Simon's poems celebrates water in some of its various forms - snow, rain, mist, dew, puddles and becks - and Pip carved most of them in situ with the help of her

Pip at work on a commemorative stone plaque in her studio

apprentice Wayne. "Two of the stones I carved in a temporary studio in a friend's barn here in Dentdale, but the rest were done on site," says Pip. "It was great on the first visit to a location, looking at the rock and then imagining it transforming from typescript into carved stone, from soft paper into a hard, massive part of the earth. There was something wonderful about being out there carving in the landscape at the mercy of the elements, seeing the views, hearing the birdsong - all that feeds into the work in a subconscious way."

The laying out of the poem on the rock, in each case, was a collaborative process - especially choosing lettering that was suitable. Because the shape of the stones affected the way the letters could appear on the rock, sometimes that meant making slight changes. "The transformation that I don't think either Simon or I expected was the way that occasionally the poem was rewritten on the rock, that the rock was somehow taking part in the writing of the poem," says Pip. "Simon was very clear from the outset that the poem was the expression and the wording had to be as clear and communicative as possible. He didn't want the lettering to get in the way of interpretation and that was a lovely creative process. My task was to devise a letterform not only sturdy enough to be visible when carved into the rock but that would act as a vessel for the poetry. It had to be sober in style and yet able to reveal the poems directly to the readers with grace and clarity, allowing their own understanding and interpretation."

Pip had to build up a kind of relationship with each stone while she was working on it, discovering new things about it as she progressed. "Every stone had its own story and character. There are aspects about each of the stones that are very special to me, to do with the material or the colour or the spirit of the place," she says. "For example, for me the Rain Stone, which is up on Blackstone Edge along the

Pennine Way, was the most dramatic location. It felt like I was on a stage because I was so high up and the ground drops away. The footpath traverses the edge and walkers would stop to watch and to hear about what I was doing. I met some lovely people up there."

Carving the Beck Stone, on Ilkley Moor, was probably the most challenging in terms of location. "I was standing in a waterfall to work," laughs Pip. "I was wearing a dry suit and I was hardly able to move but it all added to the process. Something I have learned is that gritstone has many variations - the texture, the hardness and the colour of the stones all vary enormously. Some are incredibly hard and quartzy and others are much smoother. I have so many great memories. Carving the Snow Stone was special, partly because it was the first one and visitors were taking an interest and beginning to talk about the project. It was wonderful to hear people's thoughts and to see how it captured people's imagination. The stones and poems will be there for a very long time and it is a great feeling to have been part of making this mark."

Pip was involved in a similar project - the Eden Valley Poetry Path - in 2004, commissioned by the East Cumbria Countryside Project as a response to the 2001 Foot and Mouth crisis to highlight the role of the farmers in shaping the landscape. "It was about making the countryside more accessible along existing footpaths and about interpreting the landscape through poems and natural heritage," says Pip. "There are twelve poems, written by Meg Peacocke, carved into locally-sourced stones. Each stone also has a carved picture representing an aspect of farming activity." Guide books are available in Kirkby Stephen along with paper and crayon packs for taking rubbings of the stone pictures.

Another ECCP project, Discover Eden, involved Pip in

designing and producing linocut prints for way-marking a series of walking routes in the Eden catchment, for which guide books are available. She etched a total of eighty-four bronze panels illustrating local flora and fauna, rural crafts, ancient farming life and historical events.

Not all of Pip's public art projects are in the countryside, however. In 2010, she was commissioned by Sheffield City Council's Arts Officer to work with poet Matt Black to create fifteen stone benches around the city centre related to and celebrating the city's markets. The markets in Sheffield began with a Royal Charter in 1296 and in their heyday at the end of the nineteenth century there were seven large markets in central Sheffield. "Matt created some very evocative poems for which I designed and carved many different letterforms expressing the variety and diversity of Sheffield's markets," says Pip. "We also created an alphabet trail, the Angels' Alphabet, which was made up of items that you might have bought at the markets in the 1890s."

A project currently on the easel is a commemorative stone plaque to go on the front of the house of two local historians, David and Anthea Boulton, who live in the village. "They have found out information about all the people who have lived in their house since it was built in the early seventeenth century," says Pip, who is carving the names and occupations of all the residents of the house and the dates they lived there. "It's a wonderful 'document' that seems to describe the history of the dale. You start off with wool combers and then move on to a schoolmaster - there are even stonemasons in there. Then the coal agents start appearing and later the platelayers - railway workers. It's fascinating and it's been a lovely project in terms of the lettering."

The railway was being laid at the same time that the schoolhouse where Pip lives was being built. Just after Pip bought the school, an elderly lady came to visit and explained that she had been a pupil there when she was a child. "She told me that some children used to come over here to school from Horton-in-Ribblesdale - they would come on the train." Dent station, which is twenty minutes' walk from Pip's house, is still in use, making even the small hamlet of Cowgill, apparently in the middle of nowhere, very accessible. That suits Pip who says that she is happy to go where the work is - a forthcoming public art project will require her to commute to Sheffield for eight weeks. "I like being an itinerant lettercarver. Often I am working on site and there is something wonderful about working outdoors, but I also enjoy the workshop rhythm to life. When I am in the studio, I like to listen to music while I am working - particularly Bach."

Music is one of Pip's passions - she plays the piano and the violin for relaxation and is a member of the Lakeland Fiddlers who meet every week. "It started out as an adult education course - we do gigs and play for ceilidhs and weddings, and we recently made a CD. My other loves are sketching and drawing; and I'm exploring different stone processes, such as sand-blasting. I have a passion for printmaking and lino-cutting for wallpaper: having the house to renovate now, it's a blank canvas for me to explore. I like applied art and decorative art. I recently went to an exhibition while I was in London of Cecil Collins' art - his philosophy and his art was so inspiring that when I got home I ripped up the carpet in the kitchen, which I had wanted to do for ages, and painted the floor in bright colours."

There is a small artistic community in the dale - Pip has already collaborated with Lucy Sandys-Clarke, the artisan blacksmith who works from the smithy in Dent, and would like to again - some of whom take part in the open studios schemes run by Cumbria Art Festival and the Dentdale

A view of Dentdale

Music and Beer Festival. Pip is keen to encourage others to develop new skills and regularly teaches on lettercarving courses which she now plans to offer at her house, Kirkthwaite Old School, alongside other activities. "I would like to run all sorts of residential courses here. I want it to be a place where people can come to be creative," she says. "I'd like to offer painting holidays, stone-carving and other crafts - especially those related to the landscape. It is very important to me - I draw inspiration from it. It's very subtle the way the landscape influences my designing and decision-making - sometimes it can be the sounds that I hear, often it's unconscious. I have lived here now longer than I have been in any other place and it feels a great privilege to be here."

www.piphall.co.uk

www.memorialartscharity.org.uk

14

*

Zarina Belk

Tea Room owner, Upper Wharfedale

The attractive village of Kettlewell in Upper Wharfedale has become famous in recent times due to its connection with the 2003 *Calendar Girls* movie - it was one of the filming locations - but for walkers in the area Kettlewell is perhaps better known as a stop on the long-distance walk, The Dales Way. Many a weary walker will have dropped in for a cup of tea and a bite to eat at Zarina's, a welcoming tea room in the centre of the village, run by Zarina Belk.

Zarina, who was born in Leeds, moved to Starbotton (a small hamlet between Kettlewell and Buckden) when she was eight years old. Her mother was originally from Somerset where she grew up on a farm and city life didn't really suit her. "She wanted to get away from Leeds," says Zarina. "She married my dad very young - he was a cameraman with Yorkshire Television and he went on to set up his own film company - but when I was six they split up.

My mum then met Bob who became my stepdad and I acquired another brother, Peter: I had one already, Glenn, but Bob also had a son."

She spent two years at Kettlewell Primary School but, because she enjoyed it so much there, it seems like a lot longer in Zarina's recollection. "I remember it vividly. I have very fond memories of that time - the teachers were lovely and it was a very happy school. We had a playing field which seemed massive and we could run for what seemed like miles - there was so much space. At the school I had been at in Leeds you were just a number whereas here you were a person."

Her secondary school at Threshfield, a school small enough for all the teachers to know the parents, was also a positive experience and Zarina remains an advocate for the benefits of raising children in the Dales. "People say there's nothing to do here, but there is, especially for youngsters.

Zarina's tea room in Kettlewell

You can climb trees and rocks, go paddling and swimming in the river. There haven't been many amenities but the local teenagers don't cause problems. It makes it special - people have values here. Women can go into any local pub on their own and not feel uncomfortable." Zarina feels that politicians, instead of speaking of 'broken Britain' and the lack of community feeling, could learn a lot from visiting places like Kettlewell and becoming better acquainted with the network of village life, a life based on mutual respect and looking out for each other's needs.

When her family moved to Grassington, Zarina attended Upper Wharfedale School and for the sixth form travelled to South Craven School at Sutton-in-Craven where she studied Maths, English, Child Care and Computer Studies and took RSA courses in typing and shorthand. Zarina found the academic side of school difficult as she was dyslexic - although, like many of her generation, her dyslexia went

Zarina in her tea room

undiagnosed until much later in life. However, she believes that the experience has in fact shaped her into the person she is today. "I always struggled at school and I was put down quite a lot. I think that's what made me an achiever, really - I had a lot of determination to prove them all wrong!"

After leaving school, Zarina wanted to go into hotel management and took on employment in Oxford only to find that she wasn't being trained, simply being used as a waitress. She moved on to work at the Devonshire Arms at Bolton Abbey and also at the former Wilson Arms in Threshfield, which is now a residential home. "I think looking back, that was the start of what I am doing now," she says. Her ambition was to become a restaurant manageress so, in many respects, now owning her own tea room with bed and breakfast fulfils that original ambition - and more. She enjoyed the silver service side of restaurant work but was less keen on the shifts - and the lack of a social life - so she went back to college to take a B.Tech National Diploma in Business Studies.

She went to work in Grassington at WV Patricks - a building firm - looking after the wages. With the downturn in the building trade, she was made redundant, and the same fate followed after her time at Kirk's in Skipton, where she worked in accounts. She had a temporary job in WH Smith's in Bradford as an inputter and found she liked working with computers. In due course she worked for the North Yorkshire Police Force in Skipton in the control room - but this meant working shifts again. Later, she worked for the police at Crosshills, near Keighley, and in Settle. Finally, she worked day shifts for CID administration in Skipton but then, at the age of twenty-nine, her world suddenly changed.

"I got pregnant. I had a boyfriend who was local but he was working in Scotland in the building trade. He didn't want to keep the baby, but I did - and I'm very glad I did because it was the best thing that ever happened to me."

From that moment on, Zarina took the decision to be self-supporting. Harriet was born in May 1994 and in September of that year Zarina took over the Stork Exchange in Grassington selling second-hand baby and children's clothes, toys and equipment. "People brought items in to me and I would sell them for them. I got fifty percent but most of that went on the rates and lighting. It wasn't an easy business, but it was good experience." She was determined not to claim a penny from Harriet's father, nor from the state - but the responsibility she had taken on was far from easy. Initially, it meant keeping a log of all exchange items and trying to build up a supply of all baby needs, which can be difficult in the second-hand market. As a consequence, Zarina started to buy in inexpensive new items and the business began to develop. Before too long, Zarina moved premises to a larger building in Grassington which allowed her to stock and display more goods.

This was not to be the only major change in Zarina's life. Through Harriet, Zarina came to know a little boy called Christopher and, subsequently, his father, David Belk. David was a widower whose wife had died when Christopher was only five. David was working away in London and Christopher was being looked after by his Gran. "Christopher would drag his dad into the shop on Saturday mornings and they started staying longer and longer," says Zarina. "Christopher found me, then I found his dad. It was love at first sight for me but I didn't know what to do because I knew that David had only recently lost his wife. But it all felt very natural and it happened quite quickly - David proposed to me on New Year's Eve 1999 and when we told Harriet and Christopher we were getting married they were really excited. They went around telling everyone,

The River Wharfe near Kettlewell

'We are getting married!' They have always called each other brother and sister and us mum and dad. It's not been easy putting two families together but we are an old-fashioned family and we spend a lot of time talking to each other which I think is really important." David and Zarina got married in July 2001 at Threshfield Methodist Church. "It was lovely - all the villagers were out on the village green when we came out of the church."

Three years later, on the day of Zarina's fortieth birthday, David was made redundant. He had worked for Fitzpatrick's, a construction firm responsible for, amongst other things, the pedestrianisation of Trafalgar Square. Losing his job was a blow, but commuting down to London had become, with time, more of a strain so the prospect of looking for work locally was very appealing.

When friends came north to visit one weekend, David and Zarina took them to Kettlewell and saw that a tea room was for sale. They both fell in love with the idea of taking on the business, especially as it also provided spacious accommodation. However, it took several months to make everything happen. Stork Exchange had to be sold, they needed to sell their house in Threshfield and so it was November 2004 before they could move. Nevertheless, they were both very excited about the prospect of forging a very different kind of life for themselves - and no more commuting for David.

The business they had bought was formerly The Corner Shop and Tea Room which had belonged to a local family - the Woodrups - for over a hundred years. During the first year, Zarina changed the configuration of the tea room so that the large front window offered a view for her customers. A new kitchen was installed to Zarina's design using reclaimed timber. She stripped back to the original counter and preserved the original flagstone flooring. In the refurbishment, they discovered the old fireplace where they now have a log-burning stove.

During the delay in acquiring the tea room, David saw that another business was up for sale in Kettlewell, Over and Under, which sold clothing and other equipment for outdoor pursuits. At one point it seemed that it might be one or the other business that they might run but, when the tea room was definitely available, they made a decision to do both - with Zarina in charge of the tea room and David in charge of the shop. Both enterprises are very much family businesses with Harriet helping out in the tea room and Christopher lending a hand in the shop. "It's been good for Harriet and Christopher because they have learnt how to work with people and how to run a business."

Zarina's pleasure in her work is obvious as is her desire continually to add to her customers' experience. "I love doing this and each year I keep looking for ways to make it better. One of the nicest things that people can say when they come here is that they feel like they are at home. That's the biggest compliment anyone can pay me because that's exactly how I want people to feel. There is no piped music, I serve everything in traditional china cups, saucers and plates and I don't have any sauces in little packets - mayonnaise is served in a little bowl and milk in a jug. All our food we make from fresh - some people still complain that it's slow - but I want to give people the kind of experience I would like. I don't want to eat a sandwich that was made three hours ago. I want things to taste right and not necessarily look perfect. I want people to come in and enjoy everything about it - to slow down, take time over eating and drinking and enjoy the scenery. I think it's really about wanting to give people what I think the towns have lost." Most people are very appreciative of the fact that everything is made freshly on the premises and are happy to enjoy the view from the tea room window and escape the pressures of certain aspects of modern life.

Zarina is supportive of the need to be environmentally friendly and for more people to abide by an essential principle - know what you are eating and avoid what she calls "the rubbish". She is particularly critical of the effect of supermarket chains in putting a distance between customers and an understanding of the provenance and the quality of the food they consume. "We serve proper coffee and leaf tea and local food as much as possible. I have got hens in the garden producing eggs and we grow our own vegetables. I hope the world comes back to that because a lot of the time I don't think we know what we are eating these days. It would be nice to get back to local food, to encourage people to grow their own and buy locally."

Recently, Zarina started offering bed and breakfast with two en suite double rooms available and a single room, popular with walkers who are on their own. Zarina is keen to make that room a special place to stay since it is often commented that accommodation for those by themselves seems unloved and neglected compared with rooms for couples or families. "We get quite a lot of single walkers doing the Dales Way, so it's nice to be able to offer that service."

What Zarina most enjoys about Kettlewell is that the village still has its own locally-owned shops, three pubs (when other villages can't necessarily sustain even one), a post office, a youth hostel (the post office is part of the youth hostel!) and, most importantly, a school. Famously, the villagers made it onto the pages of the national newspapers with their campaign to keep the school open, enlisting the support of best-selling author and former schools inspector Gervase Phinn, who referred to rural schools as "the central heartbeat of a village". Zarina agrees wholeheartedly. "Everyone in Kettlewell was united in that belief," she says. "If you close a school, then the families won't move here and then that's the beginning of the end for a village."

The drawback of the life Zarina has chosen is that she never really gets much of an opportunity to stand back from the business. However, now that her children are grown up and adhering to school holidays is no longer a concern, she is contemplating closing for January to allow an opportunity for a proper family holiday, perhaps abroad. In the past she has never really explored far afield - and, anyway, really admires the beauty of Upper Wharfedale. "I haven't ventured far from this area but I think I appreciate it more than some people do. Maybe it's because I came here from a city. I would rather live here and go into Leeds for certain things than live there. I love living in the countryside. You do need to have a car, though, and that's one of the reasons that the youngsters move away - and to somewhere they can afford the housing."

Zarina imagines that Kettlewell will change but, hopefully, slowly - gradual evolution is no bad thing, she feels. "You have to change and adapt to modern life but there is also a strong feeling in the village that we wouldn't want it to change too much and to try and keep what we have got - the pace of village life and a sense of community." Mutual respect is the quality of life she holds most dear, and the willingness to help out when the need arises - as with the campaign to save the school or when, recently, the snows came and the bus had to drop off the secondary school children several miles away. Everyone in the village with suitable vehicles (including quad bikes) came to the rescue.

Young families are moving in to the village - thanks, partly, to the school and its excellent reputation - as well as those couples who choose to retire here, often after holidaying in the Dales year after year. The majority of residents are those whose families have lived here for generations. What is noticeable, though, Zarina observes, is that businesses reliant on tourists, such as bed and breakfasts, quite regularly change hands. Anyone who thinks that running such a business is either easy or lucrative needs to think again, she advises.

Zarina's tea room, reliant on her commitment and long hours, is essentially seasonal, running full-time from Easter to September. Popular events, such as Kettlewell's now famous Scarecrow Festival, do bring in different customers and Zarina believes the festival is good for the village as a whole in promoting Kettlewell as a fun tourist destination. However, she has noted that her regular customers do tend to stay away at that time. There is always a risk that such events can get too big for their own good as, she feels, has happened with the Dickensian Festival each December in Grassington where the crowds of tourists now have to be bussed in using a park and ride system to regulate the numbers and traffic.

Zarina by the fireplace in her tea room

Local shops are not necessarily benefiting, she argues, when there are so many stallholders competing for business.

The village's association with the *Calendar Girls* film has also attracted tourists, while the real 'Calendar Girls' live not too far away in Rylstone and Cracoe. Zarina was an extra in the film and knows several of the original 'Girls'. In fact Ros, (who is represented as 'Ruth' in the film) worked in the tea room for a while. The 'Calendar Girls' launched the official opening of the renovated business and she has since come to know Trish and Linda. Beryl lived in Grassington and Angela (whose husband, John, died of leukaemia and was the inspiration for the subsequent calendar in order to raise funds) is a friend of Zarina's through her involvement with the amateur dramatic group the Grassington Players. The film, she believes, was a good influence for the village. "It certainly brought visitors to the Yorkshire Dales, especially from overseas, but the effect was quite short-lived and very positive." They haven't been inundated whereas, in locations associated with other Yorkshire-based television series such as *Heartbeat* and *Last of the Summer Wine*, the effect of hordes of tourists arriving all at once has been difficult for some local residents.

There is an official *Calendar Girls* walk around the village and its surroundings, and money raised from selling the leaflet has purchased a playground for Kettlewell's children. "Despite rumours to the contrary, the 'Calendar Girls' have only ever raised money for charity and have not made a penny for themselves," says Zarina who contributes herself by selling the calendars and 'Calendar Girl jam' (her own idea). Over the years she has raised more than thirteen thousand pounds for the charity Leukaemia Research. Most recently Zarina has appeared in the Grassington Players stage version of *Calendar Girls*. The Players were given special permission to perform the first ever amateur stage version in August 2012. Again, all the profits from this event were donated to John's charity. "The 'Calendar Girls' have formed a significant part of my life and my involvement in the show was a way in which I felt I could make an extra contribution," says Zarina. "They are such lovely ladies - whenever they come in to the tea room they just cheer you up and make you feel really happy. They are so close, loving and supportive and they helped Angela get through a very difficult time when John died. They are inspirational."

Happy in her marriage and with her family, as well as living in one of the country's loveliest locations, Zarina, through her own significant efforts, has made a success of her business enterprises. Her story - very much like that of *Calendar Girls* - is heart-warming, uplifting and inspirational.

www.zarinaskettlewell.co.uk

The authors

Yvette Huddleston has been a freelance journalist for twenty-two years. She was a film columnist for *The Daily Mail* for ten years and has contributed features and reviews to a variety of national and regional publications including *The Daily Telegraph, Empire* film magazine and *The Yorkshire Post*. After living in London for many years, she moved to Ilkley with her family twelve years ago. With Walter she has written *A Day in a Dale*, a guide to Yorkshire's most beautiful locations, which was published in 2011.

Walter Swan is the Business Development Manager at Ilkley Playhouse and a freelance writer. He has also enjoyed a career in English teaching and broadcasting having worked as a researcher and programme associate for the BBC, LWT and Sky. With Yvette he has been a regular contributor to *The Yorkshire Post* and *Yorkshire Ridings* magazine. A keen photographer, he has lived with his family in Burley-in-Wharfedale since 1997.

By the same authors

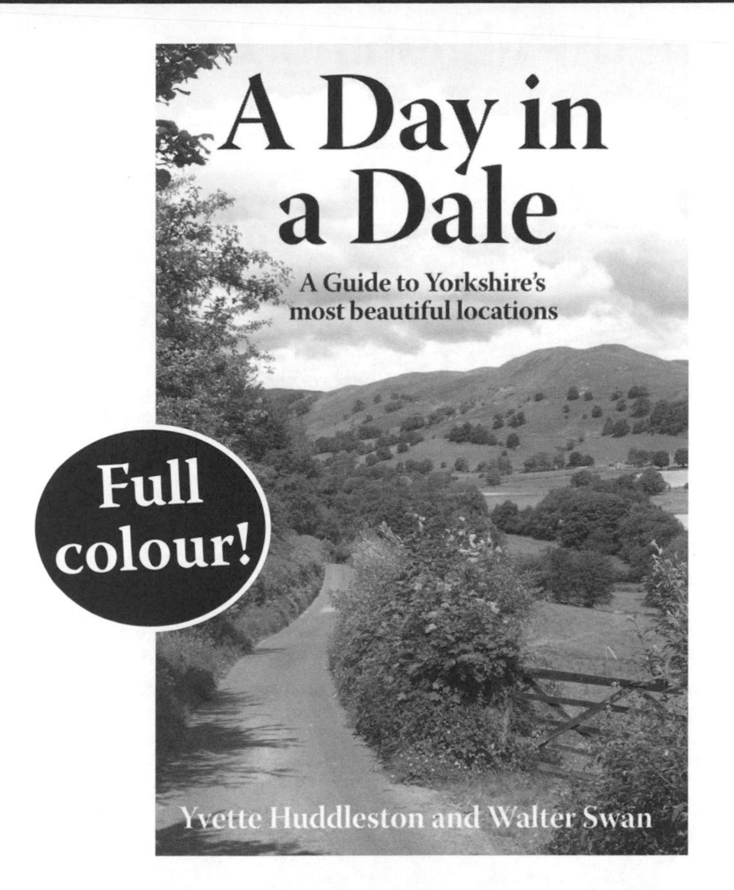

A Day in A Dale

Yvette Huddleston and Walter Swan

ISBN:978-0956478740

Scratching Shed Publishing Ltd